MATLAB Programming for Numerical Analysis

César Pérez López

Apress®

MATLAB Programming for Numerical Analysis

ISBN-13 (pbk): 978-1-4842-0296-8

ISBN-13 (electronic): 978-1-4842-0295-1

Publisher: Heinz Weinheimer
Lead Editor: Dominic Shakeshaft
Editorial Board: Steve Anglin, Mark Beckner, Ewan Buckingham, Gary Cornell, Louise Corrigan, Jim DeWolf, Jonathan Gennick, Jonathan Hassell, Robert Hutchinson, Michelle Lowman, James Markham, Matthew Moodie, Jeff Olson, Jeffrey Pepper, Douglas Pundick, Ben Renow-Clarke, Dominic Shakeshaft, Gwenan Spearing, Matt Wade, Steve Weiss
Coordinating Editor: Jill Balzano
Copy Editor: Barnaby Sheppard
Compositor: SPi Global
Indexer: SPi Global
Artist: SPi Global
Cover Designer: Anna Ishchenko

Distributed to the book trade worldwide by Springer Science+Business Media New York, 233 Spring Street, 6th Floor, New York, NY 10013. Phone 1-800-SPRINGER, fax (201) 348-4505, e-mail orders-ny@springer-sbm.com, or visit www.springeronline.com Apress Media, LLC is a California LLC and the sole member (owner) is Springer Science + Business Media Finance Inc (SSBM Finance Inc). SSBM Finance Inc is a Delaware corporation.f

For information on translations, please e-mail rights@apress.com, or visit www.apress.com.

Apress and friends of ED books may be purchased in bulk for academic, corporate, or promotional use. eBook versions and licenses are also available for most titles. For more information, reference our Special Bulk Sales–eBook Licensing web page at www.apress.com/bulk-sales.

Any source code or other supplementary material referenced by the author in this text is available to readers at www.apress.com. For detailed information about how to locate your book's source code, go to www.apress.com/source-code/.

Contents at a Glance

Contents

About the Author

César Pérez López is a Professor at the Department of Statistics and Operations Research at the University of Madrid. César is also a Mathematician and Economist at the National Statistics Institute (INE) in Madrid, a body which belongs to the Superior Systems and Information Technology Department of the Spanish Government. César also currently works at the Institute for Fiscal Studies in Madrid.

Coming Soon

- *MATLAB Differential Equations*, 978-1-4842-0311-8
- *MATLAB Control Systems Engineering*, 978-1-4842-0290-6
- *MATLAB Linear Algebra*, 978-1-4842-0323-1
- *MATLAB Differential and Integral Calculus*, 978-1-4842-0305-7
- *MATLAB Matrix Algebra*, 978-1-4842-0308-8

CHAPTER 1

■ ■ ■

The MATLAB Environment

Starting MATLAB on Windows. The MATLAB working environment

To start MATLAB, simply double-click on the shortcut icon to the program on the Windows desktop. Alternatively, if there is no desktop shortcut, the easiest and most common way to run the program is to choose *programs* from the Windows *Start* menu and select *MATLAB*. Having launched MATLAB by either of these methods, the welcome screen briefly appears, followed by the screen depicted in Figure 1-1, which provides the general environment in which the program works.

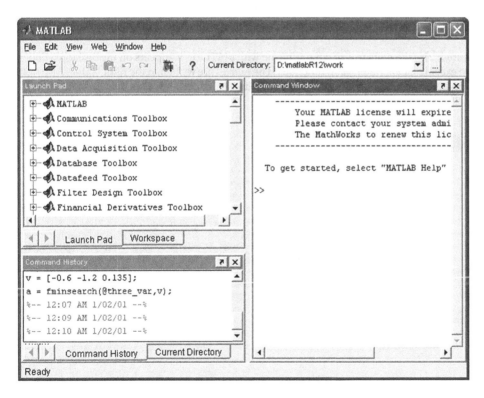

Figure 1-1.

The most important elements of the MATLAB screen are the following:

- *The Command Window*: This runs MATLAB functions.

- *The Command History*: This presents a history of the functions introduced in the Command Window and allows you to copy and execute them.

- *The Launch Pad*: This runs tools and gives you access to documentation for all MathWorks products currently installed on your computer.

- *The Current Directory*: This shows MATLAB files and execute files (such as opening and search for content operations).

- *Help (support)*: This allows you to search and read the documentation for the complete family of MATLAB products.

- *The Workspace*: This shows the present contents of the workspace and allows you to make changes to it.

- *The Array Editor*: This displays the contents of arrays in a tabular format and allows you to edit their values.

- *The Editor/Debugger*: This allows you to create, edit, and check M-files (files that contain MATLAB functions).

The MATLAB Command Window

The Command Window (Figure 1-2) is the main way to communicate with MATLAB. It appears on the desktop when MATLAB starts and is used to execute all operations and functions. The entries are written to the right of the prompt >> and, once completed, they run after pressing *Enter*. The first line of Figure 1-3 defines a matrix and, after pressing *Enter*, the matrix itself is displayed as output.

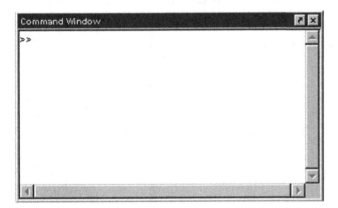

Figure 1-2.

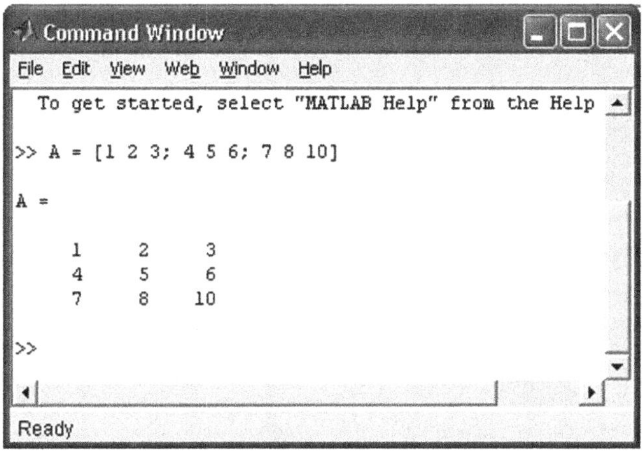

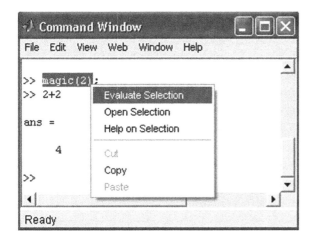

Figure 1-3.

In the Command Window, it is possible to evaluate previously executed operations. To do this, simply select the syntax you wish to evaluate, right-click, and choose the option *Evaluate Selection* from the resulting pop-up menu (Figures 1-4 and 1-5). Choosing *Open Selection* from the same menu opens in the *Editor/Debugger* an M-file previously selected in the Command Window (Figures 1-6 and 1-7).

Figure 1-4.

Figure 1-5.

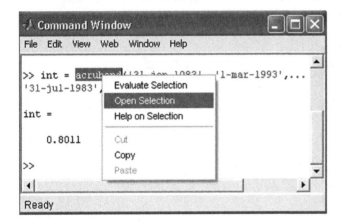

Figure 1-6.

```
function int = acrubond(id,sd,fd,rv,cpn,per,basis)
%ACRUBOND Accrued interest of security with periodic interest payments.
%        INT = ACRUBOND(ID,SD,FD,RV,CPN,PER,BASIS) returns the
%        accrued interest for a security with periodic interest payments.  This
%        function computes the accrued interest for securities with standard,
%        short, and long first coupon periods.  ID is the issue date, SD is the
%        settlement date, FD is the first coupon date, RV is the par value, CPN
%        is the coupon rate, PER is the number of periods per year (default =
%        2), and BASIS is the day-count basis: 0 = actual/actual (default), 1 =
%        30/360, 2 = actual/360, 3 = actual/365.  Enter dates as serial date
%        numbers or date strings.
%
%        For example,
%
%        int = acrubond('31-jan-1983', '1-mar-1993',...
%                          '31-jul-1983', 100, 0.1, 2, 0)
%
%        returns  int = 0.8011.
%
%        See also ACRUDISC, CFAMOUNTS, ACCRFRAC
%
%        Note: cfamounts or accrfrac is recommended when calculating
%              accrued interest beyond the first period.
```

Figure 1-7.

MATLAB is sensitive to the use of uppercase and lowercase characters, and blank spaces can be used before and after minus signs, colons and parentheses. MATLAB also allows you to write several commands on the same line, provided they are separated by semicolons (Figure 1-8). Entries are executed sequentially in the order they appear on the line. Every command which ends with a semicolon will run, but will not display its output.

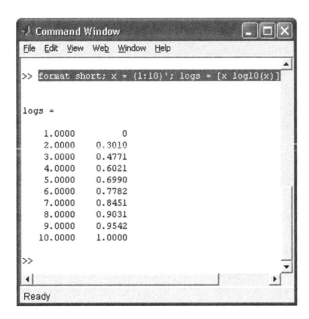

Figure 1-8.

Long entries that will not fit on one line can be continued onto a second line by placing dots at the end of the first line (Figure 1-9).

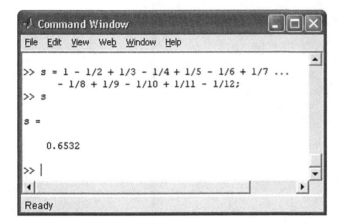

Figure 1-9.

The option *Clear Command Window* from the *Edit* menu (Figure 1-10) allows you to clear the Command Window. The command *clc* also performs this function (Figure 1-11). Similarly, the options *Clear Command History* and *Clear Workspace* in the *Edit* menu allow you to clean the history window and workspace.

Figure 1-10.

Figure 1-11.

To help you to easily identify certain elements as *if/else* instructions, chains, etc., some entries in the Command Window will appear in different colors. Some of the existing rules for colors are as follows:

1. Chains appear in purple while they are being typed. When they are finished properly (with a closing quote) they become brown.

2. Flow control syntax appears in blue. All lines between the opening and closing of the flow control functions are correctly indented.

3. Parentheses, brackets, and keys are briefly illuminated until their contents are properly completed. This allows the user to easily see if mathematical expressions are properly closed.

4. Comments in the Command Window, preceded by the symbol %, appear in green.

5. System commands such as ! appear in gold.

6. Errors are shown in red.

Below is a list of keys, arrows and combinations that can be used in the Command Window.

Key	Control key	Operation
↑	**CTRL+ p**	*Calls to the last entry submitted.*
↓	**CTRL+ n**	*Calls to the next line.*
←	**CTRL+ b**	*Moves one character backward.*
→	**CTRL+ f**	*Moves one character forward.*
CTRL+→	**CTRL+ r**	*Moves one word to the right.*
CTRL+←	**CTRL+ l**	*Moves one word to the left.*
Home	**CTRL+ a**	*Moves to the beginning of the line.*

(continued)

(*continued*)

Key	Control key	Operation
End	**CTRL+ e**	*Moves the end of the line.*
ESC	**CTRL+ u**	*Deletes the line.*
Delete	**CTRL+ d**	*Deletes the character where the cursor is.*
BACKSPACE	**CTRL+ h**	*Deletes the character before the cursor.*
	CTRL+ k	*Deletes all text up to the end of the line.*
Shift+ home		*Highlights the text from the beginning of the line.*
Shift+ end		*Highlights the text up to the end of the line.*

To enter explanatory comments simply start them with the symbol % anywhere in a line. The rest of the line should be used for the comment (see Figure 1-12).

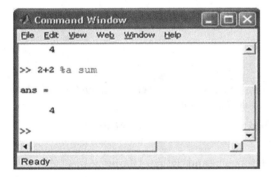

Figure 1-12.

Running M-files (files that contain MATLAB code) follows the same procedure as running any other command or function. Just type the name of the M-file (with its arguments, if necessary) in the Command Window, and press *Enter* (Figure 1-13). To see each function of an M-file as it runs, first enter the command *echo on*. To interrupt the execution of an M-file use *CTRL + c* or *CTRL + break*.

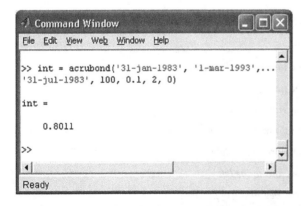

Figure 1-13.

8

Escape and exit to DOS environment commands

There are three ways to pass from the MATLAB Command Window to the MS-DOS operating system environment to run temporary assignments.

Entering the command *! dos_command* in the Command Window allows you to execute the specified command *dos_command* in the MATLAB environment. Figure 1-14 shows the execution of the command *! dir*. The same effect is achieved with the command *dos dos_command* (Figure 1-15).

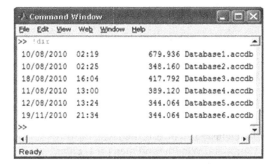

Figure 1-14.

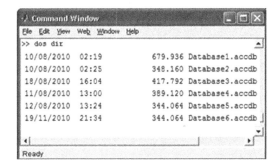

Figure 1-15.

The command *! dos_command* & is used to execute the DOS command in background mode. This opens a new window on top of the MATLAB Command Window and executes the command in that window (Figure 1-16). To return to the MATLAB environment simply click anywhere in the Command Window, or close the newly opened window via its close button ✖ or the *Exit* command.

Figure 1-16.

Not only DOS commands, but also all kinds of executable files or batch tasks can be executed with the three previous commands. To leave MATLAB simply type *quit* or *exit* in the Command Window and then press *Enter*. Alternatively you can select the option *Exit MATLAB* from the *File* menu (Figure 1-17).

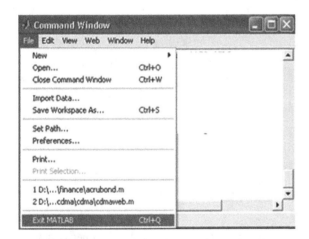

Figure 1-17.

Preferences for the Command Window

Selecting the *Preferences* option from the *File* menu (Figure 1-18) allows you to set particular features for working in the Command Window. To do this, simply choose the desired options in the *Command Window Preferences* window (Figure 1-19).

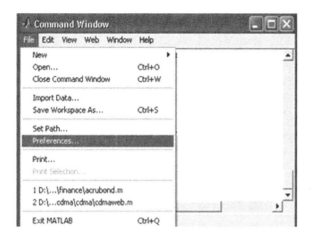

Figure 1-18.

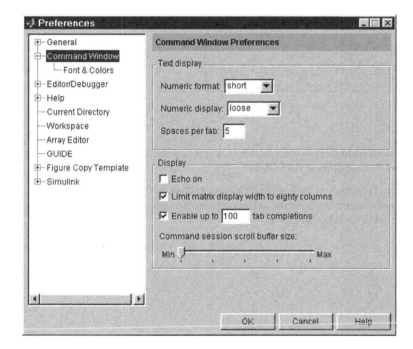

Figure 1-19.

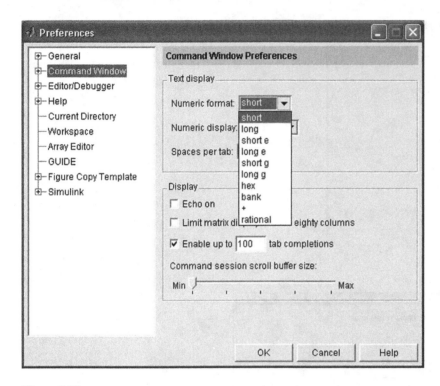

Figure 1-20.

The first area that appears in the *Command Window Preferences* window is *Text display*. This specifies how the output will appear in the Command Window. Your options are as follows:

- *Numeric format*: Specifies the format of numerical values in the Command Window (Figure 1-21).
  This affects only the appearance of the numbers, not the calculations or how to save them.
  The possible formats are presented in the following table:

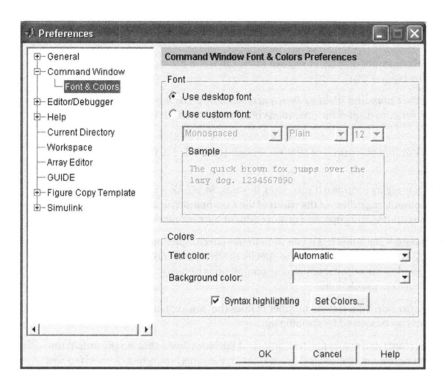

Figure 1-21.

Format	Result	Example
+	+,-, *white*	+
Bank	*Fixed*	3.14
Compact	*Removes excess lines displayed on the screen to present a more compact output.*	theta = pi/2 theta = 1.5708
Hex	*Hexadecimal*	400921fb54442d18
long	*15 digits fixed point*	3.14159265358979
long e	*15 digits floating-point*	3. 141592653589793e + 00
long g	*The best of the previous two*	3.14159265358979
loose	*Adds lines to make the output more readable. The compact command does the opposite.*	theta = pi/2 theta=1.5708
rat	*Ratio of small integers*	355/13 (a rational approximation of pi)
short	*5 digits fixed point*	3.1416
short e	*5 digits floating-point*	3. 1416e + 00
short g	*The best of the previous two*	3.1416

- *Numeric display*: Regulates the spacing of the output in the Command Window. Compact is used to suppress blank lines. Loose is used to show blank lines.

- *Spaces per tab*: Regulates the number of spaces assigned to the tab when the output is displayed (the default value is 4).

The second zone that appears in the *Command Window Preferences* window is *Display*. This specifies the size of the buffer and allows you to choose whether to display the executions of all the commands included in M-files. Your options are as follows:

- *Echo on*: If you check this box, the executions of all the commands included in the M-files are displayed.

- *Limit matrix display width to eighty columns*: If you check this box, MATLAB will display only an 80-column dot matrix output, regardless of the width of the Command Window. If this box is not checked, the matrix output will occupy the current width of the Command Window.

- *Enable up to n tab completions*: Check this box if you want to use tab completion when typing functions in the Command Window. You then need to specify the maximum number of completions that will be listed. If the number of possible completions exceeds this number, MATLAB will not show the list of completions.

- *Command session scroll buffer size*: This sets the number of lines that are kept in the Command Window buffer. These lines can be viewed by scrolling up.

In MATLAB it is also possible to set fonts and colors for the Command Window. To do this, simply unfold the sub-option *Font & Colors* hanging from *Command Windows* (Figure 1-21). In the *fonts* area select *Use desktop font* if you want to use the same source as specified for *General Font & Colors preferences*. To use a different font click the button *Use custom font* and in the three boxes located immediately below choose the desired font (Figure 1-22), style (Figure 1-23) and size. The *Sample* area shows an example of the selected font. In the *Colors* area you can choose the color of the text (*Text color*) (Figure 1-24) and the color of the background (*Background color*). If the *Syntax highlighting* box is checked, you can choose which colors will represent various types of MATLAB commands. The *Set Colors* button is used to select a given color.

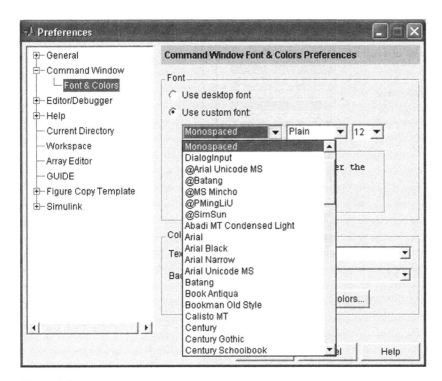

Figure 1-22.

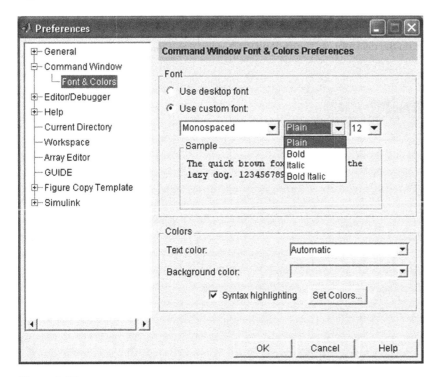

Figure 1-23.

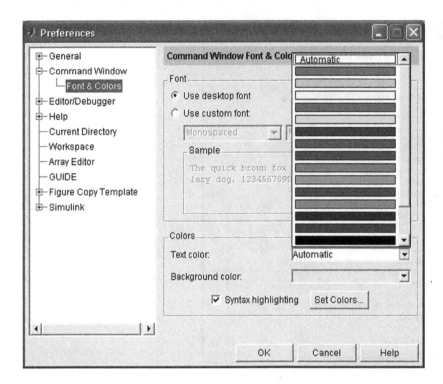

Figure 1-24.

To display the MATLAB Command Window separately simply click on the button 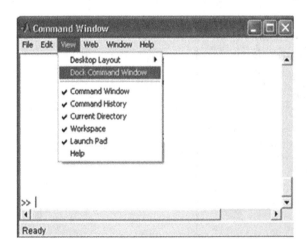 located in the top right corner. To return the window to its site on the desktop, use the option *Dock Command Window* from the *View* menu (Figure 1-25).

Figure 1-25.

The Command History window

The Command History window (Figure 1-26) appears when you start MATLAB. It is located at the bottom right of the MATLAB desktop. The Command History window shows a list of functions used recently in the Command Window (Figure 1-26). It also shows an indicator of the beginning of the session. To display this window, separated from the MATLAB desktop, simply click on the button ⤴ located in its top right corner. To return the window to its site on the desktop, use the *Dock Window Command* from the *View* menu. This method of separation and docking is common to all MATLAB windows.

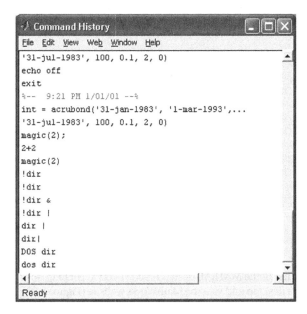

Figure 1-26.

If you select one or more lines in the Command History window and right-click on the selection, the pop-up menu of Figure 1-27 appears. This gives you options to copy the selection to the clipboard (*Copy*), evaluate the selection in the Command Window (*Evaluate Selection*), create an M-file with the selected syntax (*Create M-File*), delete the selection (*Delete Selection*), delete everything preceding the selection (*Delete to Selection*) and delete the entire history (*Delete Entire History*).

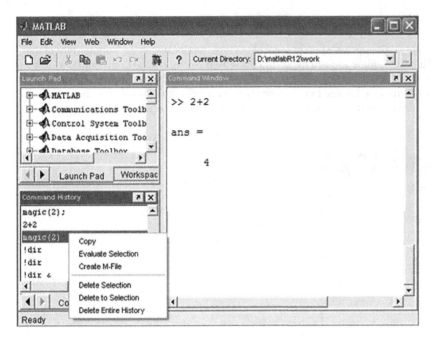

Figure 1-27.

The Launch Pad window

The Launch Pad window (located by default in the upper-left corner of the MATLAB desktop) allows you to get help, see demonstrations of installed products, go to other windows on the desktop and visit the MathWorks website (Figure 1-28).

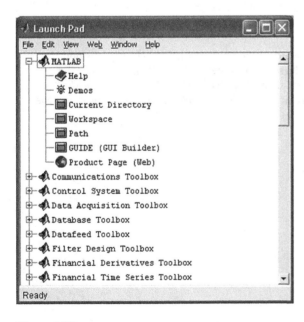

Figure 1-28.

The Current Directory window

The Current Directory window is obtained by clicking on the *Current Directory* sticker located at the bottom left of the MATLAB desktop (Figure 1-29). Its function is to view, open, and make changes in the MATLAB files environment. To display this window, separated from the MATLAB desktop (Figure 1-30), just click on the button ↗ located in its top right corner. To return the window to its site on the desktop, use the *Dock Command Window* option in the *View* menu.

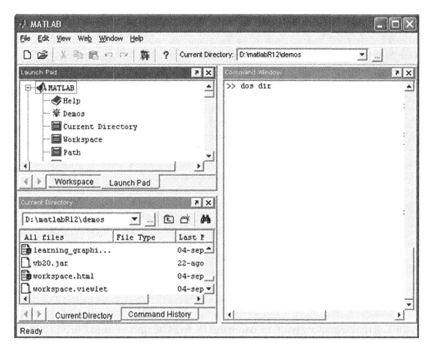

Figure 1-29.

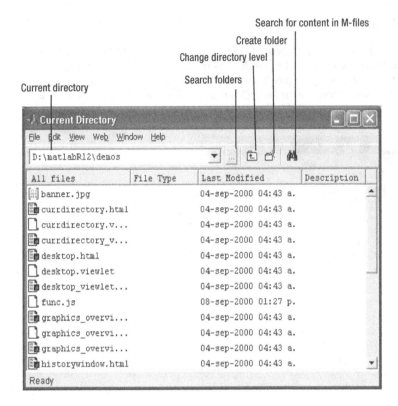

Figure 1-30.

It is possible to set preferences in the Current Directory window using the *Preferences* option from the File menu (Figure 1-31). This gives you the *Current Directory Preferences* window (Figure 1-32). In the *History* field the number of recent directories is set to save to history. In the field *Browser display options* file characteristics are set to display (file type, date of last modification, and descriptions and comments from the M-files).

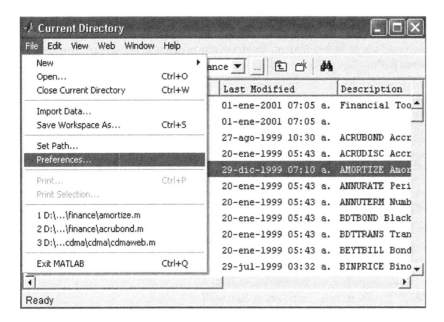

Figure 1-31.

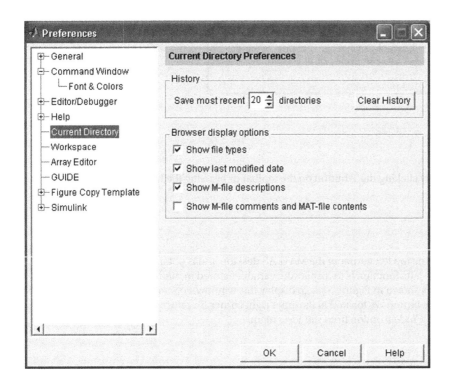

Figure 1-32.

If you select any file in the *Current Directory* window and you left-click on it, the pop-up menu of Figure 1-33 will appear. This gives you options to open the file (*Open*), run it (*Run*), view Help (*View Help*), open it as text (*Open as Text*), import data (*Import Data*), create new files, M-files or folders (*New*), rename it, delete it, cut it, copy it or paste it, pass you filters and add it to the current path.

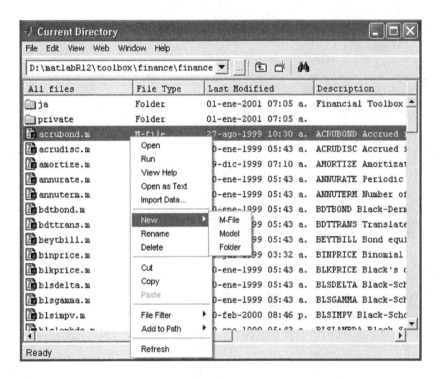

Figure 1-33.

The help browser

MATLAB's help browser is obtained by clicking the ❓ button on the toolbar or by using the function helpbrowser in the Command Window.

The Workspace window

The Workspace window is located in the top left corner of the MATLAB desktop and is obtained by clicking on the label *Work Space* under it (Figure 1-34). Its function is to display the variables stored in memory. It shows the name, type, size and class of each variable, as shown in Figure 1-35. To display this window, separated from the MATLAB desktop (Figure 1-35), just click on the button ↗ located in its upper right corner. To return the window to its site on the desktop, use the *Dock Command Window* option from the *View* menu.

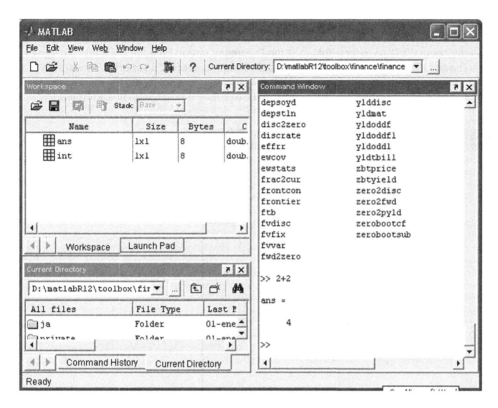

Figure 1-34.

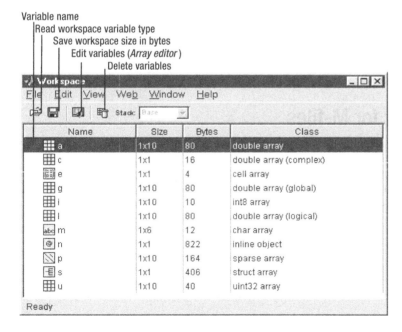

Figure 1-35.

23

An important element of the *Workspace* window is the *Array editor*, which allows you to edit numeric arrays and strings.

It is possible to set preferences in the *Workspace* window via the *Preferences* option from the *File* menu. This gives you the *Preferences* window shown in Figure 1-36. In the *History* field the number of recent directories is set to save to history. In the *Font* field the sources to be used in the Command Window preferences are set, and the option *Confirm Deletion of Variables* is checked according to whether or not you want the deletion of variables to be confirmed.

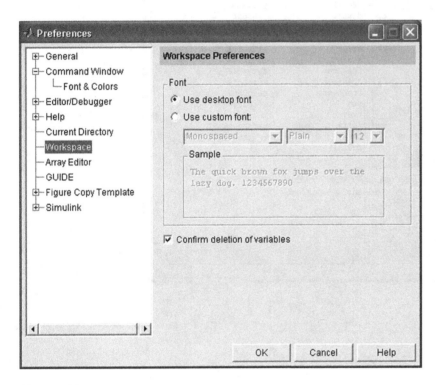

Figure 1-36.

The Editor and Debugger for M-files

To create a new M-file in the *Editor/Debugger* simply click the button ☐ in the MATLAB *Tools* toolbar or select *File* ➤ *New* ➤ *M-file* in the MATLAB desktop (Figure 1-37). The *Editor/Debugger* opens a file in which you create an M-file, i.e. a blank file for MATLAB programming code (see Figure 1-38). The *Edit* command in the Command Window also opens the *Editor/Debugger*. To open an existing M-file use *File* ➤ *Open* in the MATLAB desktop. You can also use the command *Open* in the Command Window.

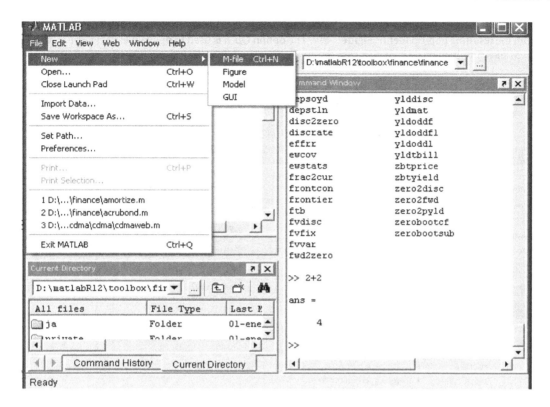

Figure 1-37.

```
function sequence=collatz(n)
% Collatz problem. Generate a sequence of integers resolving to 1
% For any positive integer, n:
%     Divide n by 2 if n is even
%     Multiply n by 3 and add 1 if n is odd
%     Repeat for the result
%     Continue until the result is 1
%
sequence = n;
next_value = n;
while next_value > 1
    if rem(next_value,2)==0
        next_value = next_value/2;
    else
        next_value = 3*next_value+1;
    end
    sequence = [sequence, next_value];
end
```

Figure 1-38.

You can also open the *Editor/Debugger* by right-clicking anywhere in the *Current Directory* window and choosing *New* ➤ *M-file* from the resulting pop-up menu (Figure 1-39). The option *Open* is used to open an existing M-file. You can open several M-files simultaneously, in which case they will appear in different windows (Figure 1-40).

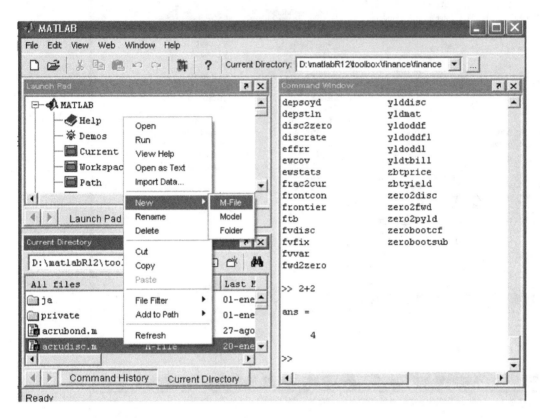

Figure 1-39.

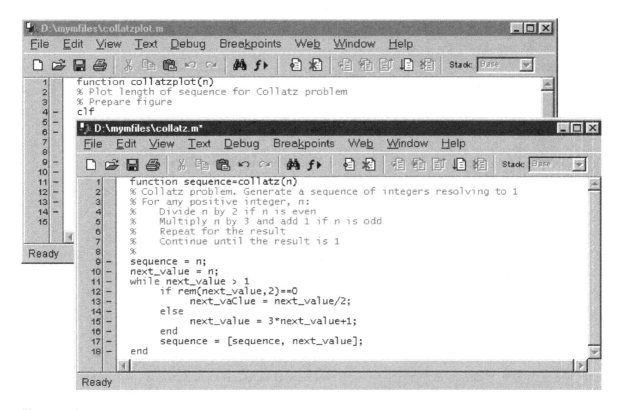

Figure 1-40.

Help in MATLAB

MATLAB has a fairly efficient inline help system. The first tool to consider is browser support (Figure 1-41), which is accessed via the icon ❓ or by typing *helpbrowser* in the Command Window (the *Help Browser* option must be selected in the *View* menu). Selecting a theme in the pane on the left of the help browser will present help on the selected topic in the right pane, and you can navigate through the content via hyperlinks. The top bar of the left navigation pane features the options *Content* (support for content), *Index* (help by alphabetical index), *Search* (find help by subject) and *Favorites* (favorite help topics).

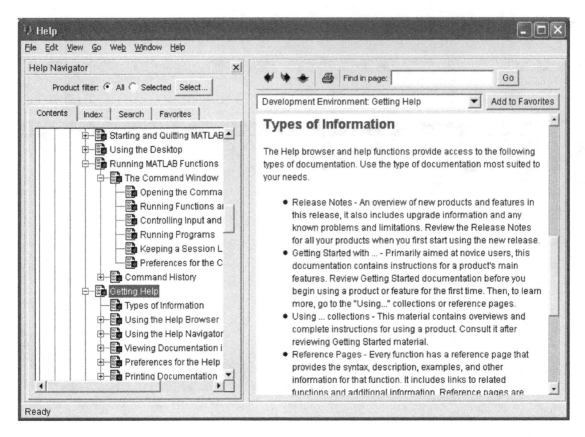

Figure 1-41.

Another very important way to obtain help in MATLAB is via its support functions. These functions are presented in the following table.

Function	Description
doc function	Displays the reference page in the browser's support for the specified function, showing syntax, description, examples and links with other related functions.
docopt	This function is used to display the location of the help files on UNIX platforms that do not support Java interfaces.
help function	Displays in the Command Window a description and the syntax of the specified function.
helpbrowser	Opens the help browser.
helpdesk	Opens the help browser. It has been replaced by doc in recent versions of MATLAB.
helpwin or helpwin theme	Displays in the help browser a list of all the MATLAB functions or those relating to the specified topic.
lookfor text	Displays in the browser all support functions which contain the specified text as part of the function.
web url	Opens in the Web browser the URL specified by default as relative to the Web help of MATLAB.

CHAPTER 2

MATLAB Language: Variables, Numbers, Operators and Functions

Variables

MATLAB does not require a command to declare variables. A variable is created simply by directly allocating a value to it. For example:

```
>> v = 3

v =

3
```

The variable v will take the value 3 and using a new mapping will not change its value. Once the variable is declared, we can use it in calculations.

```
>> v ^ 3

ans =

27

>> v+5

ans =

8
```

The value assigned to a variable remains fixed until it is explicitly changed or if the current MATLAB session is closed.

If we now write:

```
>> v = 3 + 7
```

```
v =
```

```
10
```

then the variable v has the value 10 from now on, as shown in the following calculation:

```
>> v ^ 4
```

```
ans =
```

```
10000
```

A variable name must begin with a letter followed by any number of letters, digits or underscores. However, bear in mind that MATLAB uses only the first 31 characters of the name of the variable. It is also very important to note that MATLAB is case sensitive. Therefore, a variable named with uppercase letters is different to the variable with the same name except in lowercase letters.

Vector variables

A vector variable of n elements can be defined in MATLAB in the following ways:

```
V = [v1, v2, v3,..., vn]
```

```
V = [v1 v2 v3... vn]
```

When most MATLAB commands and functions are applied to a vector variable the result is understood to be that obtained by applying the command or function to each element of the vector:

```
>> vector1 = [1,4,9,2.25,1/4]
```

```
vector1 =
```

```
1.0000 4.0000 9.0000 2.2500 0.2500
```

```
>> sqrt (vector1)
```

```
ans =
```

```
1.0000 2.0000 3.0000 1.5000 0.5000
```

The following table presents some alternative ways of defining a vector variable without explicitly bracketing all its elements together, separated by commas or blank spaces.

variable = [a:b]	*Defines the vector whose first and last elements are a and b, respectively, and the intermediate elements differ by one unit.*
variable = [a:s:b]	*Defines the vector whose first and last elements are a and b, respectively, and the intermediate elements differ by an increase specified by s.*
variable = linspace [a, b, n]	*Defines the vector with n evenly spaced elements whose first and last elements are a and b respectively.*
variable = logspace [a, b, n]	*Defines the vector with n evenly logarithmically spaced elements whose first and last elements are 10^a and 10^b, respectively.*

Below are some examples:

>> **vector2 = [5:5:25]**

vector2 =

5 10 15 20 25

This yields the numbers between 5 and 25, inclusive, separated by 5 units.

>> vector3=[10:30]

vector3 =

Columns 1 through 13

10   11   12   13   14   15   16   17   18   19   20   21   22

Columns 14 through 21

23 24 25 26 27 28 29 30

This yields the numbers between 10 and 30, inclusive, separated by a unit.

>> **t:Microsoft.WindowsMobile.DirectX.Vector4 = linspace (10,30,6)**

t:Microsoft.WindowsMobile.DirectX.Vector4 =

10 14 18 22 26 30

This yields 6 equally spaced numbers between 10 and 30, inclusive.

>> **vector5 = logspace (10,30,6)**

vector5 =

1. 0e + 030 *

0.0000 0.0000 0.0000 0.0000 0.0001 1.0000

This yields 6 evenly logarithmically spaced numbers between 10^{10} and 10^{30}, inclusive.

One can also consider row vectors and column vectors in MATLAB. A column vector is obtained by separating its elements by semicolons, or by transposing a row vector using a single quotation mark at the end of its definition.

```
>> a=[10;20;30;40]

a =

10
20
30
40

>> a=(10:14);b=a'

b =

10
11
12
13
14

>> c=(a')'

c =

10 11 12 13 14
```

You can also select an element of a vector or a subset of elements. The rules are summarized in the following table:

x (n)	*Returns the n-th element of the vector x.*
x(a:b)	*Returns the elements of the vector x between the a-th and the b-th elements, inclusive.*
x(a:p:b)	*Returns the elements of the vector x located between the a-th and the b-th elements, inclusive, but separated by p units (a > b).*
x(b:-p:a)	*Returns the elements of the vector x located between the b-th and a-th elements, both inclusive, but separated by p units and starting with the b-th element (b > a).*

Here are some examples:

```
>> x =(1:10)

x =

1    2    3    4    5    6    7    8    9    10

>> x (6)

ans =

6
```

This yields the sixth element of the vector *x*.

>> x(4:7)

ans =

4 5 6 7

This yields the elements of the vector *x* located between the fourth and seventh elements, inclusive.

>> x(2:3:9)

ans =

2 5 8

This yields the three elements of the vector *x* located between the second and ninth elements, inclusive, but separated in steps of three units.

>> x(9:-3:2)

ans =

9 6 3

This yields the three elements of the vector *x* located between the ninth and second elements, inclusive, but separated in steps of three units and starting at the ninth.

Matrix variables

MATLAB defines arrays by inserting in brackets all its row vectors separated by a comma. Vectors can be entered by separating their components by spaces or by commas, as we already know. For example, a 3 × 3 matrix variable can be entered in the following two ways:

M = [a11 a12 a13;a21 a22 a23;a31 a32 a33]

M = [a11,a12,a13;a21,a22,a23;a31,a32,a33]

Similarly we can define an array of variable dimension *(M×N)*. Once a matrix variable has been defined, MATLAB enables many ways to insert, extract, renumber, and generally manipulate its elements. The following table shows different ways to define matrix variables.

A(m,n)	*Defines the (m, n)-th element of the matrix A (row m and column n).*
A(a:b,c:d)	*Defines the subarray of A formed between the a-th and the b-th rows and between the c-th and the d-th columns, inclusive.*
A(a:p:b,c:q:d)	*Defines the subarray of A formed by every p-th row between the a-th and the b-th rows, inclusive, and every q-th column between the c-th and the d-th column, inclusive.*
A([a b],[c d])	*Defines the subarray of A formed by the intersection of the a-th through b-th rows and c-th through d-th columns, inclusive.*
A([a b c...], [e f g...])	*Defines the subarray of A formed by the intersection of rows a, b, c,... and columns e, f, g,...*
A(:,c:d)	*Defines the subarray of A formed by all the rows in A and the c-th through to the d-th columns.*
A(:,[c d e...])	*Defines the subarray of A formed by all the rows in A and columns c, d, e,...*
A(a:b,:)	*Defines the subarray of A formed by all the columns in A and the a-th through to the b-th rows.*
A([a b c...],:)	*Defines the subarray of A formed by all the columns in A and rows a, b, c,...*
A(a,:)	*Defines the a-th row of the matrix A.*
A(:,b)	*Defines the b-th column of the matrix A.*
A(:)	*Defines a column vector whose elements are the columns of A placed in order below each other.*
A(:,:)	*This is equivalent to the entire matrix A.*
[A, B, C,...]	*Defines the matrix formed by the matrices A, B, C,...*
S$_A$ = [ ]	*Clears the subarray of the matrix A, S$_A$, and returns the remainder.*
diag (v)	*Creates a diagonal matrix with the vector v in the diagonal.*
diag (A)	*Extracts the diagonal of the matrix as a column vector.*
eye (n)	*Creates the identity matrix of order n.*
eye (m, n)	*Create an m×n matrix with ones on the main diagonal and zeros elsewhere.*
zeros (m, n)	*Creates the zero matrix of order m×n.*
ones (m, n)	*Creates the matrix of order m×n with all its elements equal to 1.*
rand (m, n)	*Creates a uniform random matrix of order m×n.*
randn (m, n)	*Create a normal random matrix of order m×n.*
flipud (A)	*Returns the matrix whose rows are those of A but placed in reverse order (from top to bottom).*
fliplr (A)	*Returns the matrix whose columns are those of A but placed in reverse order (from left to right).*
rot90 (A)	*Rotates the matrix A counterclockwise by 90 degrees.*
reshape(A, m, n)	*Returns an m×n matrix formed by taking consecutive entries of A by columns.*
size (A)	*Returns the order (size) of the matrix A.*
find (cond$_A$)	*Returns all A items that meet a given condition.*
length (v)	*Returns the length of the vector v.*
tril (A)	*Returns the lower triangular part of the matrix A.*
triu (A)	*Returns the upper triangular part of the matrix A.*
A'	*Returns the transpose of the matrix A.*
Inv (A)	*Returns the inverse of the matrix A.*

Here are some examples:

We consider first the *2 × 3* matrix whose rows are the first six consecutive odd numbers:

```
>> A = [1 3 5; 7 9 11]
```

A =

```
1 3 5
7 9 11
```

Now we are going to change the *(2,3)-th* element, i.e. the last element of *A*, to zero:

```
>> A(2,3) = 0
```

A =

```
1 3 5
7 9 0
```

We now define the matrix *B* to be the transpose of *A*:

```
>> B = A'
```

B =

```
1 7
3 9
5 0
```

We now construct a matrix *C*, formed by attaching the identity matrix of order 3 to the right of the matrix *B*:

```
>> C = [B eye (3)]
```

C =

```
1   7   1   0   0
3   9   0   1   0
5   0   0   0   1
```

We are going to build a matrix *D* by extracting the odd columns of the matrix *C*, a matrix *E* formed by taking the intersection of the first two rows of *C* and its third and fifth columns, and a matrix *F* formed by taking the intersection of the first two rows and the last three columns of the matrix *C*:

```
>> D = C(:,1:2:5)
```

D =

```
1 1 0
3 0 0
5 0 1
```

```
>> E = C([1 2],[3 5])
```

E =

1 0
0 0

```
>> F = C([1 2],3:5)
```

F =

1 0 0
0 1 0

Now we build the diagonal matrix *G* such that the elements of the main diagonal are the same as those of the main diagonal of *D*:

```
>> G=diag(diag(D))
```
G =

1 0 0
0 0 0
0 0 1

We then build the matrix *H*, formed by taking the intersection of the first and third rows of *C* and its second, third and fifth columns:

```
>> H = C([1 3],[2 3 5])
```

H =

7 1 0
0 0 1

Now we build an array *I* formed by the identity matrix of order 5×4, appending the zero matrix of the same order to its right and to the right of that the unit matrix, again of the same order. Then we extract the first row of *I* and, finally, form the matrix *J* comprising the odd rows and even columns of *I* and calculate its order (size).

```
>> I = [eye(5,4) zeros(5,4) ones(5,4)]
```

ans =

1	0	0	0	0	0	0	0	1	1	1	1
0	1	0	0	0	0	0	0	1	1	1	1
0	0	1	0	0	0	0	0	1	1	1	1
0	0	0	1	0	0	0	0	1	1	1	1
0	0	0	0	0	0	0	0	1	1	1	1

```
>> I(1,:)
```

ans =

```
1    0    0    0    0    0    0    0    1    1    1    1
```

```
>> J=I(1:2:5,2:2:12)
```

J =

```
0    0    0    0    1    1
0    0    0    0    1    1
0    0    0    0    1    1
```

```
>> size(J)
```

ans =

```
3  6
```

We now construct a random matrix *K* of order *3 ×4*, reverse the order of the rows of *K*, reverse the order of the columns of *K* and then perform both operations simultaneously. Finally, we find the matrix *L* of order *4 × 3* whose columns are obtained by taking the elements of *K* sequentially by columns.

```
>> K=rand(3,4)
```

K =

```
0.5269    0.4160    0.7622    0.7361
0.0920    0.7012    0.2625    0.3282
0.6539    0.9103    0.0475    0.6326
```

```
>> K(3:-1:1,:)
```

ans =

```
0.6539    0.9103    0.0475    0.6326
0.0920    0.7012    0.2625    0.3282
0.5269    0.4160    0.7622    0.7361
```

```
>> K(:,4:-1:1)
```

ans =

```
0.7361    0.7622    0.4160    0.5269
0.3282    0.2625    0.7012    0.0920
0.6326    0.0475    0.9103    0.6539
```

```
>> K(3:-1:1,4:-1:1)

ans =

   0.6326    0.0475    0.9103    0.6539
   0.3282    0.2625    0.7012    0.0920
   0.7361    0.7622    0.4160    0.5269

>> L=reshape(K,4,3)

L =

   0.5269 0.7012 0.0475
   0.0920 0.9103 0.7361
   0.6539 0.7622 0.3282
   0.4160 0.2625 0.6326
```

Character variables

A character variable (chain) is simply a character string enclosed in single quotes that MATLAB treats as a vector form. The general syntax for character variables is as follows:

c = 'string'

Among the MATLAB commands that handle character variables we have the following:

abs ('character_string')	*Returns the array of ASCII characters equivalent to each character in the string.*
setstr (numeric_vector)	*Returns the string of ASCII characters that are equivalent to the elements of the vector.*
str2mat (t1,t2,t3,...)	*Returns the matrix whose rows are the strings t1, t2, t3,..., respectively.*
str2num ('string')	*Converts the string to its exact numeric value used by MATLAB.*
num2str (number)	*Returns the exact number in its equivalent string with fixed precision.*
int2str (integer)	*Converts the integer to a string.*
sprintf ('format', a)	*Converts a numeric array into a string in the specified format.*
sscanf ('string', 'format')	*Converts a string to a numeric value in the specified format.*
dec2hex (integer)	*Converts a decimal integer into its equivalent string in hexadecimal.*
hex2dec ('string_hex')	*Converts a hexadecimal string into its integer equivalent.*
hex2num ('string_hex')	*Converts a hexadecimal string into the equivalent IEEE floating point number.*
lower ('string')	*Converts a string to lowercase.*
upper ('string')	*Converts a string to uppercase.*
strcmp (s1, s2)	*Compares the strings s1 and s2 and returns 1 if they are equal and 0 otherwise.*
strcmp (s1, s2, n)	*Compares the strings s1 and s2 and returns 1 if their first n characters are equal and 0 otherwise.*
strrep (c, 'exp1', 'exp2')	*Replaces exp1 by exp2 in the chain c.*
findstr (c, 'exp')	*Finds where exp is in the chain c.*
isstr (expression)	*Returns 1 if the expression is a string and 0 otherwise.*

(continued)

(*continued*)

ischar (expression)	*Returns 1 if the expression is a string and 0 otherwise.*
strjust (string)	*Right justifies the string.*
blanks (n)	*Generates a string of n spaces.*
deblank (string)	*Removes blank spaces from the right of the string.*
eval (expression)	*Executes the expression, even if it is a string.*
disp ('string')	*Displays the string (or array) as has been written, and continues the MATLAB process.*
input ('string')	*Displays the string on the screen and waits for a key press to continue.*

Here are some examples:

>> **hex2dec ('3ffe56e')**

ans =

67102062

Here MATLAB has converted a hexadecimal string into a decimal number.

>> **dec2hex (1345679001)**

ans =

50356E99

The program has converted a decimal number into a hexadecimal string.

>> **sprintf(' %f',[1+sqrt(5)/2,pi])**

ans =

2.118034 3.141593

The exact numerical components of a vector have been converted to strings (with default precision).

>> **sscanf('121.00012', '%f')**

ans =

121.0001

Here a numeric string has been passed to an exact numerical format (with default precision).

>> **num2str (pi)**

ans =

3.142

39

The constant π has been converted into a string.

```
>> str2num('15/14')
```

ans =

1.0714

The string has been converted into a numeric value with default precision.

```
>> setstr(32:126)
```

ans =

!"#$% &' () * +, -. / 0123456789:; < = >? @ABCDEFGHIJKLMNOPQRSTUVWXYZ [\] ^
_'abcdefghijklmnopqrstuvwxyz {|}~

This yields the ASCII characters associated with the whole numbers between 32 and 126, inclusive.

```
>> abs('{]}><#¡¿?ªª')
```

ans =

123 93 125 62 60 35 161 191 63 186 170

This yields the integers corresponding to the ASCII characters specified in the argument of *abs*.

```
>> lower ('ABCDefgHIJ')
```

ans =

abcdefghij

The text has been converted to lowercase.

```
>> upper('abcd eFGHi jKlMn')
```

ans =

ABCD EFGHI JKLMN

The text has been converted to uppercase.

```
>> str2mat ('The world',' The country',' Daily 16', ' ABC')
```

ans =

The world
The country
Daily 16
ABC

The chains comprising the arguments of *str2mat* have been converted to a text array.

```
>> disp('This text will appear on the screen')
```

ans =

This text will appear on the screen

Here the argument of the command *disp* has been displayed on the screen.

```
>> c = 'This is a good example';
>> strrep(c, 'good', 'bad')
```

ans =

This is a bad example

The string *good* has been replaced by *bad* in the chain *c*. The following instruction locates the initial position of each occurrence of *is* within the chain *c*.

```
>> findstr (c, 'is')
```

ans =

3 6

Numbers

In MATLAB the arguments of a function can take many different forms, including different types of numbers and numerical expressions, such as integers and rational, real and complex numbers.

Arithmetic operations in MATLAB are defined according to the standard mathematical conventions. MATLAB is an interactive program that allows you to perform a simple variety of mathematical operations. MATLAB assumes the usual operations of sum, difference, product, division and power, with the usual hierarchy between them:

$x + y$	*Sum*
$x - y$	*Difference*
$x * y$ or $x\,y$	*Product*
x/y	*Division*
$x \wedge y$	*Power*

To add two numbers simply enter the first number, a plus sign (+) and the second number. Spaces may be included before and after the sign to ensure that the input is easier to read.

```
>> 2 + 3
```

ans =

5

We can perform power calculations directly.

```
>> 100 ^ 50
```

ans =

1. 0000e + 100

Unlike a calculator, when working with integers, MATLAB displays the full result even when there are more digits than would normally fit across the screen. For example, MATLAB returns the following value of $99 \wedge 50$ when using the vpa function (here to the default accuracy of 32 significant figures).

```
>> vpa '99 ^ 50'
```

ans =

. 60500606713753665044791996801256e100

To combine several operations in the same instruction one must take into account the usual priority criteria among them, which determine the order of evaluation of the expression. Consider, for example:

```
>> 2 * 3 ^ 2 + (5-2) * 3
```

ans =

27

Taking into account the priority of operators, the first expression to be evaluated is the power $3 \wedge 2$. The usual evaluation order can be altered by grouping expressions together in parentheses.

In addition to these arithmetic operators, MATLAB is equipped with a set of basic functions and you can also define your own functions. MATLAB functions and operators can be applied to symbolic constants or numbers.

One of the basic applications of MATLAB is its use in realizing arithmetic operations as if it were a conventional calculator, but with one important difference: the precision of the calculation. Operations are performed to whatever degree of precision the user desires. This unlimited precision in calculation is a feature which sets MATLAB apart from other numerical calculation programs, where the accuracy is determined by a word length inherent to the software, and cannot be modified.

The accuracy of the output of MATLAB operations can be relaxed using special approximation techniques which are exact only up to a certain specified degree of precision. MATLAB represents results with accuracy, but even if internally you are always working with exact calculations to prevent propagation of rounding errors, different approximate representation formats can be enabled, which sometimes facilitate the interpretation of the results. The commands that allow numerical approximation are the following:

format long	*Delivers results to 16 significant decimal figures.*
format short	*Delivers results to 4 decimal places. This is MATLAB's default format.*
format long e	*Provides the results to 16 decimal figures more than the power of 10 required.*
format short e	*Provides the results to four decimal figures more than the power of 10 required.*
format long g	*Provides the results in optimal long format.*
format short g	*Provides the results in optimum short format.*
bank format	*Delivers results to 2 decimal places.*
format rat	*Returns the results in the form of a rational number approximation.*
format +	*Returns the sign (+, -) and ignores the imaginary part of complex numbers.*
format hex	*Returns results in hexadecimal format.*
vpa 'operations' n	*Returns the result of the specified operations to n significant digits.*
numeric ('expr')	*Provides the value of the expression numerically approximated by the current active format.*
digits (n)	*Returns results to n significant digits.*

Using *format* gives a numerical approximation of 174/13 in the way specified after the format command:

```
>> 174/13
```

ans =

13.3846

```
>> format long; 174/13
```

ans =

13.38461538461539

```
>> format long e; 174/13
```

ans =

1.338461538461539e + 001

```
>> format short e; 174/13
```

ans =

1.3385e + 001

```
>> format long g; 174/13
```

ans =

13.3846153846154

```
>> format short g; 174/13
```

ans =

13.385

```
>> format bank; 174/13
```

ans =

13.38

```
>> format hex; 174/13
```

ans =

402ac4ec4ec4ec4f

Now we will see how the value of *sqrt (17)* can be calculated to any precision that we desire:

```
>> vpa ' 174/13 ' 10
```

ans =

13.38461538

```
>> vpa ' 174/13 ' 15
```

ans =

13.3846153846154

```
>> digits (20); vpa ' 174/13 '
```

ans =

13.384615384615384615

Integers

In MATLAB all common operations with whole numbers are exact, regardless of the size of the output. If we want the result of an operation to appear on screen to a certain number of significant figures, we use the symbolic computation command **vpa** (*variable precision arithmetic*), whose syntax we already know.

For example, the following calculates 6^400 to 450 significant figures:

```
>> '6 vpa ^ 400' 450
```

ans =

18217977168218728251394687124089371267338971528174760667459697549333959972090532700302826780076 6283
86733147959945591636745242157445605964680105495406215017704234999886990788594743994796171248406 7309
73807365248505631155692085087859428300809999273107625073394840473935055193456574397967882415119 7232
629947748581376.

The result of the operation is precise, always displaying a point at the end of the result. In this case it turns out that the answer has fewer than 450 digits anyway, so the solution is exact. If you require a smaller number of significant figures, that number can be specified and the result will be rounded accordingly. For example, calculating the above power to only 50 significant figures we have:

```
>> '6 vpa ^ 400' 50

ans =

. 18217977168218728251394687124089371267338971528175e312
```

Functions of integers and divisibility

There are several functions in MATLAB with integer arguments, the majority of which are related to divisibility. Among the most typical functions with integer arguments are the following:

rem (n, m)	*Returns the remainder of the division of n by m (also valid when n and m are real).*
sign (n)	*The sign of n (i.e. 1 if n > 0, - 1 if n < 0).*
max (n1, n2)	*The maximum of n1 and n2.*
min (n1, n2)	*The minimum of n1 and n2.*
gcd (n1, n2)	*The greatest common divisor of n1 and n2.*
lcm (n1, n2)	*The least common multiple of n1 and n2.*
factorial (n)	*n factorial (i.e. n(n-1) (n-2)...1)*
factor (n)	*Returns the prime factorization of n.*

Below are some examples.
The remainder of division of 17 by 3:

```
>> rem (17,3)

ans =

2
```

The remainder of division of 4.1 by 1.2:

```
>> rem (4.1,1.2)

ans =

0.5000
```

The remainder of division of -4.1 by 1.2:

```
>> rem(-4.1,1.2)
```

ans =

-0.5000

The greatest common divisor of 1000, 500 and 625:

```
>> gcd (1000, gcd (500,625))
```

ans =

125.00

The least common multiple of 1000, 500 and 625:

```
>> lcm (1000, lcm (500,625))
```

ans =

5000.00

Alternative bases

MATLAB allows you to work with numbers to any base, as long as the extended symbolic math *Toolbox* is available. It also allows you to express all kinds of numbers in different bases. This is implemented via the following functions:

dec2base (decimal, n_base)	*Converts the specified decimal number to the new base n_base.*
base2dec(number,b)	*Converts the given number in base b to a decimal number.*
dec2bin (decimal)	*Converts the specified decimal number to base 2 (binary).*
dec2hex (decimal)	*Converts the specified decimal number to base 16 (hexadecimal).*
bin2dec (binary)	*Converts the specified binary number to decimal.*
hex2dec (hexadecimal)	*Converts the specified base 16 number to decimal.*

Below are some examples.
Represent in base 10 the base 2 number 100101.

```
>> base2dec('100101',2)
```

ans =

37.00

Represent in base 10 the hexadecimal number FFFFAA00.

```
>> base2dec ('FFFFAA0', 16)
```

ans =

268434080.00

Represent the result of the base 16 operation FFFAA2+FF-1 in base 10.

```
>> base2dec('FFFAA2',16) + base2dec('FF',16)-1
```

ans =

16776096.00

Real numbers

As is well known, the set of real numbers is the disjoint union of the set of rational numbers and the set of irrational numbers. A rational number is a number of the form p/q, where p and q are integers. In other words, the rational numbers are those numbers that can be represented as a quotient of two integers. The way in which MATLAB treats rational numbers differs from the majority of calculators. If we ask a calculator to calculate the sum *1/2 + 1/3 + 1/4*, most will return something like *1.0833*, which is no more than an approximation of the result.

The rational numbers are ratios of integers, and MATLAB can work with them in exact mode, so the result of an arithmetic expression involving rational numbers is always given precisely as a ratio of two integers. To enable this, activate the rational format with the command *format rat*. If the reader so wishes, MATLAB can also return the results in decimal form by activating any other type of format instead (e.g. *format short* or *format long*). MATLAB evaluates the above mentioned sum in exact mode as follows:

```
>> format rat
>> 1/2 + 1/3 + 1/4
```

ans =

13/12

Unlike calculators, MATLAB ensures its operations with rational numbers are accurate by maintaining the rational numbers in the form of ratios of integers. In this way, calculations with fractions are not affected by rounding errors, which can become very serious, as evidenced by the theory of errors. Note that, once the rational format is enabled, when MATLAB adds two rational numbers the result is returned in symbolic form as a ratio of integers, and operations with rational numbers will continue to be exact until an alternative format is invoked.

A floating point number, or a number with a decimal point, is interpreted as exact if the rational format is enabled. Thus a floating point expression will be interpreted as an exact rational expression while any irrational numbers in a rational expression will be represented by an appropriate rational approximation.

```
>> format rat
>> 10/23 + 2.45/44
```

ans =

1183 / 2412

The other fundamental subset of the real numbers is the set of irrational numbers, which have always created difficulties in numerical calculation due to their special nature. The impossibility of representing an irrational number accurately in numeric mode (using the ten digits from the decimal numbering system) is the cause of most of the problems. MATLAB represents the results with an accuracy which can be set as required by the user. An irrational number, by definition, cannot be represented exactly as the ratio of two integers. If ordered to calculate the square root of 17, by default MATLAB returns the number 5.1962.

```
>> sqrt (27)
```

```
ans =
```

```
5.1962
```

MATLAB incorporates the following common irrational constants and notions:

pi	*The number $\pi = 3.1415926...$*
exp (1)	*The number $e = 2.7182818...$*
Inf	*Infinity (returned, for example, when it encounters 1/0).*
NaN	*Uncertainty (returned, for example, when it encounters 0/0).*
realmin	*Returns the smallest possible normalized floating-point number in IEEE double precision.*
realmax	*Returns the largest possible finite floating-point number in IEEE double precision.*

The following examples illustrate how MATLAB outputs these numbers and notions.

```
>> long format
>> pi
```

```
ans =
```

```
3.14159265358979
```

```
>> exp (1)
```

```
ans =
```

```
2.71828182845905
```

```
>> 1/0
```

```
Warning: Divide by zero.
```

```
ans =
```

```
Inf
```

```
>> 0/0
```

```
Warning: Divide by zero.
```

```
ans =
```

```
NaN
```

```
>> realmin
```

ans =

2. 225073858507201e-308

```
>> realmax
```

ans =

1. 797693134862316e + 308

Functions with real arguments

The disjoint union of the set of rational numbers and the set of irrational numbers is the set of real numbers. In turn, the set of rational numbers has the set of integers as a subset. All functions applicable to real numbers are also valid for integers and rational numbers. MATLAB provides a full range of predefined functions, most of which are discussed in the subsequent chapters of this book. Within the group of functions with real arguments offered by MATLAB, the following are the most important:

Trigonometric functions

Function	Inverse
sin (x)	asin (x)
cos (x)	acos (x)
tan(x)	atan(x) and atan2(y,x)
csc (x)	acsc (x)
sec (x)	asec (x)
cot (x)	acot (x)

Hyperbolic functions

Function	Inverse
sinh (x)	asinh (x)
cosh(x)	acosh(x)
tanh(x)	atanh(x)
csch(x)	acsch(x)
sech(x)	asech(x)
coth (x)	acoth (x)

Exponential and logarithmic functions

Function	Meaning
exp (x)	*Exponential function in base e (e ^ x).*
log (x)	*Base e logarithm of x.*
log10 (x)	*Base 10 logarithm of x.*
log2 (x)	*Base 2 logarithm of x.*
pow2 (x)	*2 raised to the power x.*
sqrt (x)	*The square root of x.*

Numeric variable-specific functions

Function	Meaning
abs (x)	*The absolute value of x.*
floor (x)	*The largest integer less than or equal to x.*
ceil (x)	*The smaller integer greater than or equal to x.*
round (x)	*The closest integer to x.*
fix (x)	*Removes the fractional part of x.*
rem (a, b)	*Returns the remainder of the division of a by b.*
sign (x)	*Returns the sign of x (1 if x > 0,0 if x=0,- 1 if x < 0).*

Here are some examples:

>> sin(pi/2)

ans =

1

>> asin (1)

ans =

1.57079632679490

>> log (exp (1) ^ 3)

ans =

3.00000000000000

The function *round* is demonstrated in the following two examples:

```
>> round (2.574)
```

ans =

3

```
>> round (2.4)
```

ans =

2

The function *ceil* is demonstrated in the following two examples:

```
>> ceil (4.2)
```

ans =

5

```
>> ceil (4.8)
```

ans =

5

The function *floor* is demonstrated in the following two examples:

```
>> floor (4.2)
```

ans =

4

```
>> floor (4.8)
```

ans =

4

The *fix* function simply removes the fractional part of a real number:

```
>> fix (5.789)
```

ans =

5

Complex numbers

Operations on complex numbers are well implemented in MATLAB. MATLAB follows the convention that i or j represents the key value in complex analysis, the *imaginary number* $\sqrt{-1}$. All the usual arithmetic operators can be applied to complex numbers, and there are also some specific functions which have complex arguments. Both the real and the imaginary part of a complex number can be a real number or a symbolic constant, and operations with them are always performed in exact mode, unless otherwise instructed or necessary, in which case an approximation of the result is returned. As the imaginary unit is represented by the symbol i or j, the complex numbers are expressed in the form $a+bi$ or $a+bj$. Note that you don't need to use the product symbol (asterisk) before the imaginary unit:

```
>> (1-5i)*(1-i)/(-1+2i)

ans =

-1.6000 + 2.8000i

>> format rat
>> (1-5i) *(1-i) /(-1+2i)

ans =

-8/5 + 14/5i
```

Functions with complex arguments

Working with complex variables is very important in mathematical analysis and its many applications in engineering. MATLAB implements not only the usual arithmetic operations with complex numbers, but also various complex functions. The most important functions are listed below.

Trigonometric functions

Function	Inverse
sin (z)	*asin (z)*
cos (z)	*acos (z)*
tan (z)	*atan(z) and atan2(imag(z),real(z))*
csc (z)	*acsc (z)*
sec (z)	*asec (z)*
cot (z)	*acot (z)*

Hyperbolic functions

Function	Inverse
sinh (z)	*asinh (z)*
cosh(z)	*acosh(z)*
tanh(z)	*atanh(z)*
csch(z)	*acsch(z)*
sech(z)	*asech(z)*
coth (z)	*acoth (z)*

Exponential and logarithmic functions

Function	Meaning
exp (z)	*Exponential function in base e (e ^ z)*
log (z)	*Base e logarithm of z.*
log10 (z)	*Base 10 logarithm of z.*
log2 (z)	*Base 2 logarithm of z.*
pow2 (z)	*2 to the power z.*
sqrt (z)	*The square root of z.*

Specific functions for the real and imaginary part

Function	Meaning
floor (z)	Applies the floor function to real(z) and imag(z).
ceil (z)	Applies the ceil function to real(z) and imag(z).
round (z)	Applies the round function to real(z) and imag(z).
fix (z)	Applies the fix function to real(z) and imag(z).

Specific functions for complex numbers

Function	Meaning
abs (z)	The complex modulus of z.
angle (z)	The argument of z.
conj (z)	The complex conjugate of z.
real (z)	The real part of z.
imag (z)	The imaginary part of z.

Below are some examples of operations with complex numbers.

```
>> round(1.5-3.4i)
```

ans =

2   -   3i

```
>> real(i^i)
```

ans =

0.2079

```
>> (2+2i)^2/(-3-3*sqrt(3)*i)^90
```

ans =

0502e-085 - 1 + 7. 4042e-070i

```
>> sin (1 + i)
```

ans =

1.2985 + 0. 6350i

Elementary functions that support complex vector arguments

MATLAB easily handles vector and matrix calculus. Indeed, its name, *MAtrix LABoratory*, already gives an idea of its power in working with vectors and matrices. MATLAB allows you to work with functions of a complex variable, but in addition this variable can even be a vector or a matrix. Below is a table of functions with complex vector arguments.

max (V)	*The maximum component of V. (max is calculated for complex vectors as the complex number with the largest complex modulus (magnitude), computed with max(abs(V)). Then it computes the largest phase angle with max(angle(x)), if necessary.)*
min (V)	*The minimum component of V. (min is calculated for complex vectors as the complex number with the smallest complex modulus (magnitude), computed with min(abs(A)). Then it computes the smallest phase angle with min(angle(x)), if necessary.)*
mean (V)	*Average of the components of V.*
median (V)	*Median of the components of V.*
std (V)	*Standard deviation of the components of V.*
sort (V)	*Sorts the components of V in ascending order. For complex entries the order is by absolute value and argument.*
sum (V)	*Returns the sum of the components of V.*

(*continued*)

(continued)

prod (V)	*Returns the product of the components of V, so, for example, n! = prod(1:n).*
cumsum (V)	*Gives the cumulative sums of the components of V.*
cumprod (V)	*Gives the cumulative products of the components of V.*
diff (V)	*Gives the vector of first differences of V (V_t - V_{t-1}).*
gradient (V)	*Gives the gradient of V.*
del2 (V)	*Gives the Laplacian of V (5-point discrete).*
fft (V)	*Gives the discrete Fourier transform of V.*
fft2 (V)	*Gives the two-dimensional discrete Fourier transform of V.*
ifft (V)	*Gives the inverse discrete Fourier transform of V.*
ifft2 (V)	*Gives the inverse two-dimensional discrete Fourier transform of V.*

These functions also support a complex matrix as an argument, in which case the result is a vector of column vectors whose components are the results of applying the function to each column of the matrix.

Here are some examples:

```
>> V = 2:7, W = [5 + 3i 2-i 4i]

V =

2    3    4    5    6    7

W =

2.0000 - 1.0000i      0 + 4.0000i   5.0000 + 3.0000i

>> diff(V),diff(W)

ans =

1    1    1    1    1

ans =

-2.0000 + 5.0000i    5.0000 - 1.0000i

>> cumprod(V),cumsum(V)

ans =

2           6          24         120         720        5040

ans =

2    5    9    14    20    27
```

```
>> cumsum(W), mean(W), std(W), sort(W), sum(W)

ans =

2.0000 - 1.0000i    2.0000 + 3.0000i    7.0000 + 6.0000i

ans =

2.3333 + 2.0000i

ans =

3.6515

ans =

2.0000 - 1.0000i    0 + 4.0000i    5.0000 + 3.0000i

ans =

7.0000 + 6.0000i

>> mean(V), std(V), sort(V), sum(V)

ans =

4.5000

ans =

1.8708

ans =

2    3    4    5    6    7

ans =

27

>> fft(W), ifft(W), fft2(W)

ans =

7.0000 + 6.0000i    0.3660 - 0.1699i   -1.3660 - 8.8301i

ans =

2.3333 + 2.0000i   -0.4553 - 2.9434i    0.1220 - 0.0566i

ans =

7.0000 + 6. 0000i 0.3660 - 0. 1699i   -1.3660 - 8. 8301i
```

Elementary functions that support complex matrix arguments

Trigonometric

sin (z)	*Sine function*
sinh (z)	*Hyperbolic sine function*
asin (z)	*Arcsine function*
asinh (z)	*Hyperbolic arcsine function*
cos (z)	*Cosine function*
cosh (z)	*Hyperbolic cosine function*
acos (z)	*Arccosine function*
acosh (z)	*Hyperbolic arccosine function*
tan(z)	*Tangent function*
tanh (z)	*Hyperbolic tangent function*
atan (z)	*Arctangent function*
atan2 (z)	*Fourth quadrant arctangent function*
atanh (z)	*Hyperbolic arctangent function*
sec (z)	*Secant function*
sech (z)	*Hyperbolic secant function*
asec (z)	*Arccosecant function*
asech (z)	*Hyperbolic arccosecant function*
csc (z)	*Cosecant function*
csch (z)	*Hyperbolic cosecant function*
acsc (z)	*Arccosecant function*
acsch (z)	*Hyperbolic arccosecant function*
cot (z)	*Cotangent function*
coth (z)	*Hyperbolic cotangent function*
acot (z)	*Arccotangent function*
acoth (z)	*Hyperbolic arccotangent function*

Exponential

exp (z)	*Base e exponential function*
log (z)	*Natural logarithm function (base e)*
log10 (z)	*Base 10 logarithm function*
sqrt (z)	*Square root function*

(continued)

(continued)

Complex	
abs (z)	*Modulus or absolute value*
angle (z)	*Argument*
conj (z)	*Complex conjugate*
imag (z)	*Imaginary part*
real (z)	*Real part*
Numerical	
fix (z)	*Removes the fractional part*
floor (z)	*Rounds to the nearest lower integer*
ceil (z)	*Rounds to the nearest greater integer*
round (z)	*Performs common rounding*
rem (z1, z2)	*Returns the remainder of the division of z1 by z2*
sign (z)	*The sign of z*
Matrix	
expm (Z)	*Matrix exponential function by default*
expm1 (Z)	*Matrix exponential function in M-file*
expm2 (Z)	*Matrix exponential function via Taylor series*
expm3 (Z)	*Matrix exponential function via eigenvalues*
logm (Z)	*Logarithmic matrix function*
sqrtm (Z)	*Matrix square root function*
funm(Z,'function')	*Applies the function to the array Z*

Here are some examples:

```
>> A=[7 8 9; 1 2 3; 4 5 6], B=[1+2i 3+i;4+i,i]

A =

7    8    9
1    2    3
4    5    6

B =

1.0000 + 2.0000i   3.0000 + 1.0000i
4.0000 + 1.0000i        0 + 1.0000i
```

```
>> sin(A), sin(B), exp(A), exp(B), log(B), sqrt(B)

ans =

0.6570     0.9894     0.4121
0.8415     0.9093     0.1411
-0.7568   -0.9589    -0.2794

ans =

3.1658  + 1.9596i   0.2178 - 1.1634i
-1.1678 - 0.7682i        0 + 1.1752i

ans =

1.0e+003 *

1.0966     2.9810     8.1031
0.0027     0.0074     0.0201
0.0546     0.1484     0.4034

ans =

-1.1312 + 2.4717i   10.8523 +16.9014i
29.4995 +45.9428i    0.5403 + 0.8415i

ans =

0.8047 + 1.1071i    1.1513 + 0.3218i
1.4166 + 0.2450i         0 + 1.5708i

ans =

1.2720 + 0.7862i    1.7553 + 0.2848i
2.0153 + 0.2481i    0.7071 + 0.7071i
```

The exponential functions, square root and logarithm used above apply to the array elementwise and have nothing to do with the matrix exponential and logarithmic functions that are used below.

```
>> expm(B), logm(A), abs(B), imag(B)

ans =

-27.9191 + 14.8698i -20.0011 + 12.0638i
-24.7950 + 17.6831i -17.5059 + 14.0445i
```

```
ans =

11.9650    12.8038 -19.9093
-21.7328 -22.1157   44.6052
11.8921    12.1200 -21.2040

ans =

2.2361 3.1623
4.1231 1.0000

ans =

2    1
1    1
```

```
>> fix(sin(B)), ceil(log(A)), sign(B), rem(A,3*ones(3))

ans =

3.0000 + 1.0000i        0 - 1.0000i
-1.0000                 0 + 1.0000i

ans =

2    3    3
0    1    2
2    2    2

ans =

0.4472 + 0.8944i   0.9487 + 0.3162i
0.9701 + 0.2425i        0 + 1.0000i

ans =

1    2    0
1    2    0
1    2    0
```

Random numbers

MATLAB can easily generate (pseudo) random numbers. The function *rand* generates uniformly distributed random numbers and the function *randn* generates normally distributed random numbers. The most interesting features of MATLAB's random number generator are presented in the following table.

rand	*Returns a uniformly distributed random decimal number from the interval [0,1].*
rand (n)	*Returns an array of size n×n whose elements are uniformly distributed random decimal numbers from the interval [0,1].*
rand (m, n)	*Returns an array of dimension m×n whose elements are uniformly distributed random decimal numbers from the interval [0,1].*
rand (size (a))	*Returns an array of the same size as the matrix A and whose elements are uniformly distributed random decimal numbers from the interval [0,1].*
rand ('seed')	*Returns the current value of the uniform random number generator seed.*
rand('seed',n)	*Assigns to n the current value of the uniform random number generator seed.*
randn	*Returns a normally distributed random decimal number (mean 0 and variance 1).*
randn (n)	*Returns an array of dimension n×n whose elements are normally distributed random decimal numbers (mean 0 and variance 1).*
randn (m, n)	*Returns an array of dimension m×n whose elements are normally distributed random decimal numbers (mean 0 and variance 1).*
randn (size (a))	*Returns an array of the same size as the matrix A and whose elements are normally distributed random decimal numbers (mean 0 and variance 1).*
randn ('seed')	*Returns the current value of the normal random number generator seed.*
randn('seed',n)	*Assigns to n the current value of the uniform random number generator seed.*

Here are some examples:

```
>> [rand, rand (1), randn, randn (1)]

ans =

0.9501    0.2311   -0.4326   -1.6656
```

```
>> [rand(2), randn(2)]

ans =

0.6068    0.8913            0.1253   -1.1465
0.4860    0.7621            0.2877    1.1909
```

```
>> [rand(2,3), randn(2,3)]

ans =

0.3529 0.0099 0.2028 -0.1364 1.0668 -0.0956
0.8132 0.1389 0.1987  0.1139 0.0593 -0.8323
```

Operators

MATLAB features arithmetic, logical, relational, conditional and structural operators.

Arithmetic operators

There are two types of arithmetic operators in MATLAB: matrix arithmetic operators, which are governed by the rules of linear algebra, and arithmetic operators on vectors, which are performed elementwise. The operators involved are presented in the following table.

Operator	Role played
+	*Sum of scalars, vectors, or matrices*
-	*Subtraction of scalars, vectors, or matrices*
*	*Product of scalars or arrays*
.*	*Product of scalars or vectors*
\	$A \backslash B = inv(A) * B$, *where A and B are matrices*
.\	$A. \backslash B = [B(i,j) / A(i,j)]$, *where A and B are vectors [dim (A) = dim (B)]*
/	*Quotient, or* $B/A = B * inv(A)$, *where A and B are matrices*
./	$A / B = [A(i,j)/b(i,j)]$, *where A and B are vectors [dim (A) = dim (B)]*
^	*Power of a scalar or matrix* (M^p)
.^	*Power of vectors* $(A.\, {}^\wedge B = [A(i,j)^{B(i,j)}]$, *for vectors A and B)*

Simple mathematical operations between scalars and vectors apply the scalar to all elements of the vector according to the defined operation, and simple operators between vectors are performed element by element. Below is the specification of these operators:

a = {a1, a2,..., an}, b = {b1, b2,..., bn}, c = scalar	
a + c = [a1 +c, a2+ c,..., an+c]	Sum of a scalar and a vector
a * c = [a1 * c,a2* c ,..., an * c]	Product of a scalar and a vector
a + b = [a1+b1 a2+b2 ... an+bn]	Sum of two vectors
a. * b = [a1*b1 a2*b2 ... an*bn]	Product of two vectors
a. / b = [a1/b1 a2/b2 ... an/bn]	Ratio to the right of two vectors
a. \ b = [a1\b1 a2\b2 ... an\bn]	Ratio to the left of two vectors
a. ^ c = [a1 ^c, a2^ c ,..., an ^ c]	Vector to the power of a scalar
c. ^ a = [c ^ a1,c ^ a2,... ,c ^ an]	Scalar to the power of a vector
a.^b = [a1^b1 a2^b2 ... an^bn]	Vector to the power of a vector

It must be borne in mind that the vectors must be of the same length and that in the product, quotient and power the first operand must be followed by a point.

The following example involves all of the above operators.

```
>> X = [5,4,3]; Y = [1,2,7]; a = X + Y, b = X-Y, c = x * Y, d = 2. * X,...
e = 2/X, f = 2. \Y, g = x / Y, h =. \X, i = x ^ 2, j = 2. ^ X, k = X. ^ Y

a =

6      6      10

b =

4      2      -4

c =

5      8      21

d =

10      8      6

e =

0.4000      0.5000      0.6667

f =

0.5000      1.0000      3.5000

g =

5.0000      2.0000      0.4286

h =

5.0000      2.0000      0.4286

i =

25      16      9

j =

32 16 8

k =

5 16 2187
```

The above operations are all valid since in all cases the variable operands are of the same dimension, so the operations are successfully carried out element by element. For the sum and the difference there is no distinction between vectors and matrices, as the operations are identical in both cases.

The most important operators for matrix variables are specified below:

A + B, A - B, A * B	*Addition, subtraction and product of matrices.*
A\B	*If A is square, A\B = inv (A) * B. If A is not square, A\B is the solution, in the sense of least-squares, of the system AX = B.*
B/A	*Coincides with (A ' \ B')'.*
Aⁿ	*Coincides with A * A * A *... *A n times (n integer).*
p^A	*Performs the power operation only if p is a scalar.*

Here are some examples:

```
>> X = [5,4,3]; Y = [1,2,7]; l = X'* Y, m = X * Y ', n = 2 * X, o = X / Y, p = Y\X

l =

5 10 35
4  8 28
3  6 21

m =

34

n =

10 8 6

o =

0.6296

p =

0          0          0
0          0          0
0.7143     0.5714     0.4286
```

All of the above matrix operations are well defined since the dimensions of the operands are compatible in every case. We must not forget that a vector is a particular case of matrix, but to operate with it in matrix form (not element by element), it is necessary to respect the rules of dimensionality for matrix operations. For example, the vector operations $X. '*Y$ and $X.*Y'$ make no sense, since they involve vectors of different dimensions. Similarly, the matrix operations $X*Y$, $2/X$, $2\backslash Y$, $X \wedge 2$, $2 \wedge X$ and $X \wedge Y$ make no sense, again because of a conflict of dimensions in the arrays.

Here are some more examples of matrix operators.

```
>> M = [1,2,3;1,0,2;7,8,9]

M =

1 2 3
1 0 2
7 8 9
```

```
>> B = inv (M), C = M ^ 2, D = M ^(1/2), E = 2 ^ M
```

B =

```
-0.8889     0.3333     0.2222
 0.2778    -0.6667     0.0556
 0.4444     0.3333    -0.1111
```

C =

```
24    26    34
15    18    21
78    86    118
```

D =

```
0.5219 + 0.8432i    0.5793 - 0.0664i    0.7756 - 0.2344i
0.3270 + 0.0207i    0.3630 + 1.0650i    0.4859 - 0.2012i
1.7848 - 0.5828i    1.9811 - 0.7508i    2.6524 + 0.3080i
```

E =

```
1. 0e + 003 *
```

```
0.8626 0.9568 1.2811
0.5401 0.5999 0.8027
2.9482 3.2725 4.3816
```

Relational operators

MATLAB also provides relational operators. Relational operators perform element by element comparisons between two matrices and return an array of the same size whose elements are zero if the corresponding relationship is true, or one if the corresponding relation is false. The relational operators can also compare scalars with vectors or matrices, in which case the scalar is compared to all the elements of the array. Below is a table of these operators.

<	*Less than (for complex numbers this applies only to the real parts)*
< =	*Less than or equal (only applies to real parts of complex numbers)*
>	*Greater than (only applies to real parts of complex numbers)*
> =	*Greater than or equal (only applies to real parts of complex numbers)*
x == y	*Equality (also applies to complex numbers)*
x ~ = y	*Inequality (also applies to complex numbers)*

Logical operators

MATLAB provides symbols to denote logical operators. The logical operators shown in the following table offer a way to combine or negate relational expressions.

~ A	*Logical negation (NOT) or the complement of A.*
A & B	*Logical conjunction (AND) or the intersection of A and B.*
A | B	*Logical disjunction (OR) or the union of A and B.*
XOR (A, B)	*Exclusive OR (XOR) or the symmetric difference of A and B (takes the value 1 if A or B, but not both, are 1).*

Here are some examples:

```
>> A = 2:7;P =(A>3) &(A<6)

P =

0    0    1    1    0    0
```

Returns 1 when the corresponding element of *A* is greater than 3 and less than 6, and returns 0 otherwise.

```
>> X = 3 * ones (3.3); X > = [7 8 9; 4 5 6 ; 1 2 3]

ans =

0 0 0
0 0 0
1 1 1
```

The elements of the solution array corresponding to those elements of *X* which are greater than or equal to the equivalent entry of the matrix *[7 8 9; 456 ; 1 2 3]* are assigned the value 1. The remaining elements are assigned the value 0.

Logical functions

MATLAB implements logical functions whose output can take the value true (1) or false (0). The following table shows the most important logical functions.

exist(A)	*Checks if the variable or function exists (returns 0 if A does not exist and a number between 1 and 5, depending on the type, if it does exist).*
any(V)	*Returns 0 if all elements of the vector V are null and returns 1 if some element of V is non-zero.*
any(A)	*Returns 0 for each column of the matrix A with all null elements and returns 1 for each column of the matrix A which has non-null elements.*
all(V)	*Returns 1 if all the elements of the vector V are non-null and returns 0 if some element of V is null.*
all(A)	*Returns 1 for each column of the matrix A with all non-null elements and returns 0 for each column of the matrix A with at least one null element.*

(continued)

(*continued*)

find (V)	*Returns the places (or indices) occupied by the non-null elements of the vector V.*
isnan (V)	*Returns 1 for the elements of V that are indeterminate and returns 0 for those that are not.*
isinf (V)	*Returns 1 for the elements of V that are infinite and returns 0 for those that are not.*
isfinite (V)	*Returns 1 for the elements of V that are finite and returns 0 for those that are not.*
isempty (A)	*Returns 1 if A is an empty array and returns 0 otherwise (an empty array is an array such that one of its dimensions is 0).*
issparse (A)	*Returns 1 if A is a sparse matrix and returns 0 otherwise.*
isreal (V)	*Returns 1 if all the elements of V are real and 0 otherwise.*
isprime (V)	*Returns 1 for all elements of V that are prime and returns 0 for all elements of V that are not prime.*
islogical (V)	*Returns 1 if V is a logical vector and 0 otherwise.*
isnumeric (V)	*Returns 1 if V is a numeric vector and 0 otherwise.*
ishold	*Returns 1 if the properties of the current graph are retained for the next graph and only new elements will be added and 0 otherwise.*
isieee	*Returns 1 if the computer is capable of IEEE standard operations.*
isstr (S)	*Returns 1 if S is a string and 0 otherwise.*
ischart (S)	*Returns 1 if S is a string and 0 otherwise.*
isglobal (A)	*Returns 1 if A is a global variable and 0 otherwise.*
isletter (S)	*Returns 1 if S is a letter of the alphabet and 0 otherwise.*
isequal (A, B)	*Returns 1 if the matrices or vectors A and B are equal, and 0 otherwise.*
ismember(V, W)	*Returns 1 for every element of V which is in W and 0 for every element V that is not in W.*

Below are some examples using the above defined logical functions.

```
>> V=[1,2,3,4,5,6,7,8,9], isprime(V), isnumeric(V), all(V), any(V)

V =

1    2    3    4    5    6    7    8    9

ans =

0    1    1    0    1    0    1    0    0

ans =

1

ans =

1
```

```
ans =

1

>> B=[Inf, -Inf, pi, NaN], isinf(B), isfinite(B), isnan(B), isreal(B)

B =

Inf - Inf 3.1416 NaN

ans =

1 1 0 0

ans =

0 0 1 0

ans =

0 0 0 1

ans =

1


>> ismember ([1,2,3], [8,12,1,3]), A = [2,0,1];B = [4,0,2]; isequal (2A * B)

ans =

1 0 1

ans =

1
```

EXERCISE 2-1

Find the number of ways of choosing 12 elements from 30 without repetition, the remainder of the division of 2^{134} by 3, the prime decomposition of 18900, the factorial of 200 and the smallest number N which when divided by 16,24,30 and 32 leaves remainder 5.

```
>> factorial (30) / (factorial (12) * factorial(30-12))

ans =

8.6493e + 007
```

The command vpa is used to present the exact result.

>> vpa 'factorial (30) / (factorial (12) * factorial(30-12))' 15

ans =

86493225.

>> rem(2^134,3)

ans =

0

>> factor (18900)

ans =

2   2   3   3   3   5   5   7

>> factorial (100)

ans =

9. 3326e + 157

The command vpa is used to present the exact result.

>> vpa ' factorial (100)' 160

ans =

93326215443944152681699238856266700490715968264381621468592963895217599993229915608941463976156518286253697920827223758251185210916864000000000000000000000000.

N-5 is the least common multiple of 16, 24, 30 and 32.

>> lcm (lcm (16.24), lcm (30,32))

ans =

480

Then N = 480 + 5 = 485.

EXERCISE 2-2

In base 5 find the result of the operation defined by $a25aaff6_{16} + 6789aba_{12} + 35671_8 + 1100221_3 - 1250$. In base 13 find the result of the operation $(666551_7)^* (aa199800a_{11}) + (fffaaa125_{16})/(33331_4 + 6)$.

The result of the first operation in base 10 is calculated as follows:

```
>> base2dec('a25aaf6',16) + base2dec('6789aba',12) +...
base2dec('35671',8) + base2dec('1100221',3)-1250
```

ans =

190096544

We then convert this to base 5:

```
>> dec2base (190096544,5)
```

ans =

342131042134

Thus, the final result of the first operation in base 5 is 342131042134.

The result of the second operation in base 10 is calculated as follows:

```
>> base2dec('666551',7) * base2dec('aa199800a',11) +...
79 * base2dec('fffaaa125',16) / (base2dec ('33331', 4) + 6)
```

ans =

2. 7537e + 014

We now transform the result obtained into base 13.

```
>> dec2base (275373340490852,13)
```

ans =

BA867963C1496

EXERCISE 2-3

In base 13, find the result of the following operation:

$(666551_7)* (aa199800a_{11}) + (fffaaa125_{16}) / (33331_4 + 6)$.

First, we perform the operation in base 10:

A more direct way of doing all of the above is:

```
>> base2dec('666551',7) * base2dec('aa199800a',11) +...
79 * base2dec('fffaaa125',16) / (base2dec ('33331', 4) + 6)
```

ans =

2. 753733404908515e + 014

We now transform the result obtained into base 13.

```
>> dec2base (275373340490852,13)
```

ans =

BA867963C1496

EXERCISE 2-4

Given the complex numbers X = 2 + 2i and Y=-3-3sqrt(3)i, calculate Y^3X^2/Y^{90}, $Y^{1/2}$, $Y^{3/2}$ and ln (X).

```
>> X=2+2*i; Y=-3-3*sqrt(3)*i;
>> Y^3
```

ans =

216

```
>> X ^ 2 / Y ^ 90
```

ans =

050180953422426e-085 - 1 + 7. 404188256695968e-070i

```
>> sqrt (Y)
```

ans =

1.22474487139159 - 2.12132034355964i

```
>> sqrt(Y^3)
```

ans =

14.69693845669907

>> log (X)

ans =

1.03972077083992 + 0.78539816339745i

EXERCISE 2-5

Calculate the value of the following operations with complex numbers:

$$\frac{i^8-i^{-8}}{3-4i}+1,\ i^{\sin(1+i)},\ (2+\mathrm{In}(i))^{\frac{1}{i}},\ (1+i)^i,\ i^{\ln(1+i)},\ (1+\sqrt{3})i^{1-i}$$

>> (i^8-i^(-8))/(3-4*i) + 1

ans =

1

>> i^(sin(1+i))

ans =

-0.16665202215166 + 0.32904139450307i

>> (2+log(i))^(1/i)

ans =

1.15809185259777 - 1.56388053989023i

>> (1+i)^i

ans =

0.42882900629437 + 0.15487175246425i

>> i^(log(1+i))

ans =

0.24911518828716 + 0.15081974484717i

```
>> (1+sqrt(3)*i)^(1-i)
```

ans =

5.34581479196611 + 1. 975948834528731

EXERCISE 2-6

Calculate the real part, imaginary part, modulus and argument of each of the following expressions:

$$i^{3i}, (1+\sqrt{3})i^{1-i}, i^{i^{i}}, i^{i}$$

```
>> Z1 = i ^ 3 * i; Z2 = (1 + sqrt (3) * i) ^(1-i); Z3 =(i^i) ^ i;Z4 = i ^ i;
```

```
>> format short
```

```
>> real ([Z1 Z2 Z3 Z4])
ans =
```

1.0000 5.3458 0.0000 0.2079

```
>> imag ([Z1 Z2 Z3 Z4])
```

ans =

0 1.9759 - 1.0000 0

```
>> abs ([Z1 Z2 Z3 Z4])
```

ans =

1.0000 5.6993 1.0000 0.2079

```
>> angle ([Z1 Z2 Z3 Z4])
```

ans =

0 0.3541 - 1.5708 0

EXERCISE 2-7

Generate a square matrix of order 4 whose elements are uniformly distributed random numbers from [0,1].
Generate another square matrix of order 4 whose elements are normally distributed random numbers from [0,1].
Find the present generating seeds, change their value to ½ and rebuild the two arrays of random numbers.

>> rand (4)

ans =

```
0.9501 0.8913 0.8214 0.9218
0.2311 0.7621 0.4447 0.7382
0.6068 0.4565 0.6154 0.1763
0.4860 0.0185 0.7919 0.4057
```

>> randn (4)

ans =

```
-0.4326 -1.1465  0.3273 -0.5883
-1.6656  1.1909  0.1746  2.1832
 0.1253  1.1892 -0.1867 -0.1364
 0.2877 -0.0376  0.7258  0.1139
```

>> rand ('seed')

ans =

931316785

>> randn ('seed')

ans =

931316785

>> randn ('seed', 1/2)
>> rand ('seed', 1/2)
>> rand (4)

ans =

```
0.2190 0.9347 0.0346 0.0077
0.0470 0.3835 0.0535 0.3834
0.6789 0.5194 0.5297 0.0668
0.6793 0.8310 0.6711 0.4175
```

```
>> randn (4)

ans =

1.1650 -0.6965  0.2641  1.2460
0.6268  1.6961  0.8717 -0.6390
0.0751  0.0591 -1.4462  0.5774
0.3516  1.7971 -0.7012 -0.3600
```

EXERCISE 2-8

Given the vector variables a = [π, 2π, 3π, 4π, 5π] and b = [e, 2e, 3e, 4e, 5e], calculate c = sin (a) + b, d = cos (a), e = ln (b), f = c * d, g = c/d, h = d ^ 2, i = d ^ 2-e ^ 2 and j = 3d ^ 3-2e ^ 2.

```
>> a = [pi, 2 * pi, 3 * pi, 4 * pi, 5 * pi],
b = [exp (1), 2 * exp (1), 3 * exp (1), 4 * exp (1),5*exp(1)],
c=sin(a)+b,d=cos(a),e=log(b),f=c.*d,g=c./d,]
h=d.^2, i=d.^2-e.^2, j=3*d.^3-2*e.^2

a =

3.1416    6.2832    9.4248    12.5664    15.7080

b =

2.7183 5.4366 8.1548 10.8731 13.5914

c =

2.7183 5.4366 8.1548 10.8731 13.5914

d =

-1     1     -1     1     -1

e =

1.0000 1.6931 2.0986 2.3863 2.6094

f =

-2.7183 5.4366 - 8.1548 10.8731 - 13.5914

g =

-2.7183 5.4366 - 8.1548 10.8731 - 13.5914

h =

1     1     1     1     1
```

```
i =

0 - 1.8667 - 3.4042 - 4.6944 - 5.8092

j =

-5.0000 - 2.7335 - 11.8083 - 8.3888 - 16.6183
```

EXERCISE 2-9

Given a uniform random square matrix M of order 3, obtain its inverse, its transpose and its diagonal. Transform it into a lower triangular matrix (replacing the upper triangular entries by 0) and rotate it 90 degrees counterclockwise. Find the sum of the elements in the first row and the sum of the diagonal elements. Extract the subarray whose diagonal elements are at $_{11}$ and $_{22}$ and also remove the subarray whose diagonal elements are at $_{11}$ and $_{33}$.

```
>> M=rand(3)

M =

    0.6868    0.8462    0.6539
    0.5890    0.5269    0.4160
    0.9304    0.0920    0.7012

>> A=inv(M)

A =

   -4.1588    6.6947   -0.0934
    0.3255    1.5930   -1.2487
    5.4758   -9.0924    1.7138

>> B=M'

B =

    0.6868    0.5890    0.9304
    0.8462    0.5269    0.0920
    0.6539    0.4160    0.7012

>> V=diag(M)

V =

    0.6868
    0.5269
    0.7012
```

```
>> TI=tril(M)
```

TI =

```
0.6868    0         0
0.5890    0.5269    0
0.9304    0.0920    0.7012
```

```
>> TS=triu(M)
```

TS =

```
0.6868    0.8462    0.6539
0         0.5269    0.4160
0         0         0.7012
```

```
>> TR=rot90(M)
```

TR =

```
0.6539    0.4160    0.7012
0.8462    0.5269    0.0920
0.6868    0.5890    0.9304
```

```
>> s=M(1,1)+M(1,2)+M(1,3)
```

s =

2.1869

```
>> sd=M(1,1)+M(2,2)+M(3,3)
```

sd =

1.9149

```
>> SM=M(1:2,1:2)
```

SM =

```
0.6868 0.8462
0.5890 0.5269
```

```
>> SM1 = M([1 3], [1 3])
```

SM1 =

```
0.6868 0.6539
0.9304 0.7012
```

EXERCISE 2-10

Given the following complex square matrix M of order 3, find its square, its square root and its base 2 and − 2 exponential:

$$M = \begin{bmatrix} i & 2i & 3i \\ 4i & 5i & 6i \\ 7i & 8i & 9i \end{bmatrix}.$$

```
>> M=[i 2*i 3*i; 4*i 5*i 6*i; 7*i 8*i 9*i]

M =

0 + 1.0000i        0 + 2.0000i        0 + 3.0000i
0 + 4.0000i        0 + 5.0000i        0 + 6.0000i
0 + 7.0000i        0 + 8.0000i        0 + 9.0000i

>> C=M^2

C =

-30    -36    -42
-66    -81    -96
-102   -126   -150

>> D=M^(1/2)
D =

0.8570 - 0.2210i    0.5370 + 0.2445i    0.2169 + 0.7101i
0.7797 + 0.6607i    0.9011 + 0.8688i    1.0224 + 1.0769i
0.7024 + 1.5424i    1.2651 + 1.4930i    1.8279 + 1.4437i

>> 2^M

ans =

0.7020 - 0.6146i    -0.1693 - 0.2723i    -0.0407 + 0.0699i
-0.2320 - 0.3055i    0.7366 - 0.3220i    -0.2947 - 0.3386i
-0.1661 + 0.0036i    -0.3574 - 0.3717i    0.4513 - 0.7471i

>> (-2)^M

ans =

17.3946 -16.8443i     4.3404 - 4.5696i    -7.7139 + 7.7050i
1.5685 - 1.8595i      1.1826 - 0.5045i    -1.2033 + 0.8506i
-13.2575 +13.1252i    -3.9751 + 3.5607i    6.3073 - 6.0038i
```

EXERCISE 2-11

Given the complex matrix M in the previous exercise, find its elementwise logarithm and its elementwise base e exponential. Also calculate the results of the matrix operations e^M and ln (M).

```
>> M=[i 2*i 3*i; 4*i 5*i 6*i; 7*i 8*i 9*i]
```

```
>> log(M)
```

ans =

```
0 + 1.5708i       0.6931 + 1.5708i    1.0986 + 1.5708i
1.3863 + 1.5708i  1.6094 + 1.5708i    1.7918 + 1.5708i
1.9459 + 1.5708i  2.0794 + 1.5708i    2.1972 + 1.5708i
```

```
>> exp(M)
```

ans =

```
0.5403  + 0.8415i  -0.4161 + 0.9093i  -0.9900 + 0.1411i
-0.6536 - 0.7568i   0.2837 - 0.9589i   0.9602 - 0.2794i
0.7539  + 0.6570i  -0.1455 + 0.9894i  -0.9111 + 0.4121i
```

```
>> logm(M)
```

ans =

```
-5.4033 - 0.8472i    11.9931 - 0.3109i   -5.3770 + 0.8846i
12.3029 + 0.0537i   -22.3087 + 0.8953i   12.6127 + 0.4183i
-4.7574 + 1.6138i    12.9225 + 0.7828i   -4.1641 + 0.6112i
```

```
>> expm(M)
```

ans =

```
0.3802 - 0.6928i   -0.3738 - 0.2306i   -0.1278 + 0.2316i
-0.5312 - 0.1724i   0.3901 - 0.1434i   -0.6886 - 0.1143i
-0.4426 + 0.3479i  -0.8460 - 0.0561i   -0.2493 - 0.4602i
```

79

EXERCISE 2-12

Given the complex vector V = [1 + i, i, 1-i], find the mean, median, standard deviation, variance, sum, product, maximum and minimum of its elements, as well as its gradient, its discrete Fourier transform and its inverse discrete Fourier transform.

```
>> [mean(V),median(V),std(V),var(V),sum(V),prod(V),max(V),min(V)]'

ans =

0.6667 - 0.3333i
1.0000 + 1.0000i
1.2910
1.6667
2.0000 - 1.0000i
0 - 2.0000i
1.0000 + 1.0000i
0 - 1.0000i

>> gradient(V)

ans =

1.0000 - 2.0000i    0.5000    0 + 2.0000i

>> fft(V)

ans =

2.0000 + 1.0000i   -2.7321 + 1.0000i    0.7321 + 1.0000i

>> ifft(V)

ans =

0.6667 + 0. 3333i 0.2440 + 0. 3333i - 0.9107 + 0. 3333i
```

EXERCISE 2-13

Given the arrays

$$A = \begin{bmatrix} 1 & 1 & 0 \\ 0 & 1 & 1 \\ 0 & 0 & 1 \end{bmatrix}, B = \begin{bmatrix} i & 1-i & 2+i \\ 0 & -1 & 3-1 \\ 0 & 0 & -i \end{bmatrix}, C = \begin{bmatrix} 1 & 1 & 1 \\ 0 & sqrt(2)i & -sqrt(2)i \\ 1 & -1 & -1 \end{bmatrix}$$

calculate AB – BA, A² + B² + C², ABC, sqrt (A)+sqrt(B)+sqrt(C), $e^A(e^B + e^C)$, their transposes and their inverses. Also verify that the product of any of the matrices A, B, C with its inverse yields the identity matrix.

```
>> A=[1 1 0;0 1 1;0 0 1]; B=[i 1-i 2+i;0 -1 3-i;0 0 -i]; C=[1 1 1; 0 sqrt(2)*i -sqrt(2)*i;1
-1 -1];
```

```
>> M1=A*B-B*A
```

M1 =

```
0              -1.0000 - 1.0000i    2.0000
0               0                   1.0000 - 1.0000i
0               0                   0
```

```
>> M2=A^2+B^2+C^2
```

M2 =

```
2.0000         2.0000 + 3.4142i    3.0000 - 5.4142i
0 - 1.4142i   -0.0000 + 1.4142i    0.0000 - 0.5858i
0              2.0000 - 1.4142i    2.0000 + 1.4142i
```

```
>> M3=A*B*C
```

M3 =

```
5.0000 + 1.0000i   -3.5858 + 1.0000i   -6.4142 + 1.0000i
3.0000 - 2.0000i   -3.0000 + 0.5858i   -3.0000 + 3.4142i
0 - 1.0000i         0 + 1.0000i         0 + 1.0000i
```

```
>> M4=sqrtm(A)+sqrtm(B)-sqrtm(C)
```

M4 =

```
0.6356 + 0.8361i   -0.3250 - 0.8204i    3.0734 + 1.2896i
0.1582 - 0.1521i    0.0896 + 0.5702i    3.3029 - 1.8025i
-0.3740 - 0.2654i   0.7472 + 0.3370i    1.2255 + 0.1048i
```

```
>> M5=expm(A)*(expm(B)+expm(C))
```

M5 =

```
14.1906 - 0.0822i    5.4400 + 4.2724i   17.9169 - 9.5842i
4.5854 - 1.4972i     0.6830 + 2.1575i    8.5597 - 7.6573i
3.5528 + 0.3560i     0.1008 - 0.7488i    3.2433 - 1.8406i
```

```
>> inv(A)

ans =

1 -1  1
0 -1 -1
0  0  1

>> inv(B)

ans =

0 - 1.0000i  -1.0000 - 1.0000i  -4.0000 + 3.0000i
0              -1.0000            1.0000 + 3.0000i
0               0                 0 + 1.0000i

>> inv(C)

ans =

0.5000               0              0.5000
0.2500      0 - 0.3536i  -0.2500
0.2500      0 + 0.3536i  -0.2500

>> [A*inv(A) B*inv(B) C*inv(C)]

ans =

1   0   0   1   0   0   1   0   0
0   1   0   0   1   0   0   1   0
0   0   1   0   0   1   0   0   1

>> A'

ans =

1 0 0
1 1 0
0 1 1

>> B'

ans =

0 - 1.0000i        0                  0
1.0000 + 1.0000i  -1.0000             0
2.0000 - 1.0000i   3.0000 + 1.0000i   0 + 1.0000i

>> C'

ans =

1.0000   0              1.0000
1.0000   0 - 1.4142i  -1.0000
1.0000   0 + 1.4142i  -1.0000
```

■ ■ ■

CHAPTER 3

MATLAB Language: Development Environment Features

General Purpose Commands

MATLAB has a group of so-called general purpose commands that can be further classified into the following subcategories according to the essential function of the script:

- Commands that handle variables in the workspace.

- Commands that work with files and the operating environment.

- Command handling functions.

- Commands that control the *Command Window.*

- Commands that start and exit MATLAB.

Commands that Handle Variables in the Workspace

MATLAB allows you to define and manage variables, and store them in files, in a very simple way. When extensive calculations are performed, it is convenient to give names to intermediate results. These intermediate results are assigned to variables to make them easier to use. The definition of variables has already been treated in the previous chapter, but it is convenient to recall that the value assigned to a variable is permanent, until it is explicitly changed or the current MATLAB session is closed.

The following table presents a group of MATLAB commands that handle variables:

clear	*Clears all variables in the workspace.*
clear(v1,v2, …, vn)	*Deletes the specified numeric variables.*
clear('v1', 'v2', …, 'vn')	*Clears the specified string variables.*
disp(X)	*Displays an array without including its name.*
length(X)	*Shows the length of the vector X and if X is an array, displays its greatest dimension.*

(continued)

load	*Reads all variables from the file MATLAB.mat.*
load file	*Reads all variables specified in the .mat file.*
load file X Y Z	*Reads the variables X, Y, Z from the specified .mat file.*
load file -ascii	*Reads the file as ASCII whatever its extension.*
load file -mat	*Reads the file as .mat whatever its extension.*
S = load(…)	*Assigns the contents of a .mas file to the variable S.*
memory	*Displays how much memory is available and how much is currently being used.*
mlock	*Prevents the deletion of M-files in memory.*
munlock	*Allows the deletion of M-files in memory.*
openvar('v')	*Opens the variable v in the workspace in the Array Editor, allowing graphical editing.*
pack	*Compresses the workspace memory.*
pack file	*Used as a temporary file to store the variables.*
pack 'file'	*Functional form of pack.*
save	*Saves the variables in the workspace in the binary file MATLAB.mat in the current directory.*
save file	*Saves the variables in the workspace in the file file.mat in the current directory. A .mat file has a specific MATLAB format.*
save file v1 v2 …	*Saves the variables v1, v2,…in the workspace in the file file.mat.*
save … option	*Saves the variables in the workspace in the format specified by option.*
save('file', …)	*Functional form of save.*
saveas(h, 'f.ext')	*Saves the figure or model h as an f.ext file.*
saveas(h,'f','format')	*Saves the figure or model h as f in the specified format file.*
d = size(X)	*Returns the sizes of each dimension of an array X in a vector d.*
[m,n] = size(X)	*Returns the dimensions of the matrix X as two variables named m and n.*
[d1,d2,d3,…,dn] = size(X)	*Returns the dimensions of the array X as variables named d1, d2,…, dn.*
who	*Lists the variables in the workspace.*
whos	*Lists the variables in the workspace with sizes and types.*
who('global')	*Lists the variables in the global workspace.*
whos('global')	*Lists the variables in the global workspace with sizes and types.*
who('-file','filename')	*Lists the variables in the specified .mat file.*
whos('-file','filename')	*Lists the variables in the specified .mat file and their sizes and types.*
who('var1','var2',…)	*Lists the string variables from the specified workspace.*
who('-file','filename',	*Lists the specified string variables in the given .mat file.*
'var1','var2',…)	*Stores the list of variables in s.*
s = who(…)	*Stores the list of variables with their sizes and types in s.*
s = whos(…)	*Lists the numerical variables specified in the given .mat file.*
who -file filename var1 var2 …	*Lists the numerical variables specified in the file .mat given with their sizes and types.*
whos -file filename var1 var2…	
workspace	*Opens a browser to manage the workspace.*

The *save* command, which applies to file workspace variables, supports the following options:

Option	Mode of Storage of the Data
-append	*The variables are added to the end of the file.*
-ascii	*The variables are stored in a file in 8 digit ASCII format.*
-ascii - double	*The variables are stored in a file in 16 digit ASCII format.*
-ascii - tabs	*The variables are stored in a tab-delimited file in 8 digit ASCII format.*
-ascii - double - tabs	*The variables are stored in a tab-delimited file in 16 digit ASCII format.*
-mat	*The variables are stored in a file in binary .mat MATLAB MAT-file format.*
-v4	*The variables are stored in a file with MATLAB version 4.*

The command *save* is the essential instrument for storing data in MATLAB type.*mat* files (only readable by the MATLAB program) and ASCII type files (readable by any application). By default, variables are stored in .*mat* formatted files. To store variables in ASCII formatted files it is necessary to use options.

As a first example we let a variable *A* be equal to the inverse of a random square matrix of order 5 and a variable *B* be equal to the inverse of twice the unit matrix of order 5 less the identity matrix of order 5.

```
>> A=inv(rand(3))

A =
  1.67        -0.12        -0.93
 -0.42         1.17         0.20
 -0.85        -1.00         1.71
```

```
>> B=inv(2*ones(3)-eye(3))

B =
-0.60  0.40  0.40
 0.40 -0.60  0.40
 0.40  0.40 -0.60
```

Now we use the commands *who* and *whos* to view the workspace variables as, respectively, a simple list and a list together with types and sizes.

```
>> who

Your variables are:

A  B
```

```
>> whos

Name       Size       Bytes  Class

A          3x3          72    double array
B          3x3          72    double array

Grand total is 18 elements using 144 bytes
```

If we want only the variable information about *A*, we do the following:

```
>> who A
```

Your variables are:

A

```
>> whos A
```

Name   Size   Bytes   Class

A   3x3   72   double array

Grand total is 9 elements using 72 bytes

Now we are going to store the variables *A* and *B* in an ASCII file with 8 digits of precision and name it *matrix.asc*. In addition, to check the ASCII file has been generated, we use the command *dir* to see that our file exists. Finally, we will check the contents of our file, using the DOS operating system order *type* to check that the contents are indeed the elements of two arrays with 8 digits of precision, located one after the other.

```
>> save matrix.asc A B - ascii
>> dir
```

.   ..   matrix.asc

```
>> type matrix.asc
```

```
 1. 6740445e + 000 - 1. 1964440e-001 - 9. 2759516e-001
-4 1647244e-001 1. 1737582e + 000 2. 0499870e-001
5035677e-001 - 8 - 1. 0006147e + 000 1. 7125190e + 000
-6 0000000e-001 4. 0000000e-001 4. 0000000e-001
4. 0000000e-001 - 6. 0000000e-001 4. 0000000e-001
4. 0000000e-001 4. 0000000e-001 - 6. 0000000e-001
```

The files generated with the command *save* are stored by default (if not specified otherwise) in the \MATLAB\BIN\ subdirectory.

Saving all variables in the workspace with the command *save* to a binary file in MATLAB format is equivalent to selecting the option *Save Workspace As* from the general MATLAB *file* menu.

Once the variables have been saved, the workspace can be deleted by using the command *clear*.

```
>> clear
```

Then, to illustrate the command *load*, we will read the previously saved ASCII file *matrix.asc*. MATLAB will read the ASCII file as a single variable whose name is that of the file, as is checked with the command *whos*.

```
>> load matrix.asc
>> whos
```

```
Name            Size          Bytes  Class

matrix          6x3             144  double array

Grand total is 18 elements using 144 bytes
```

We now check that MATLAB has read the data in the same 6 × 3 matrix structure that it had been saved in, the first three rows corresponding to the variable *A* and the last three to the variable *B*.

>> matrix

```
matrix =

  1.67  -0.12  -0.93
 -0.42   1.17   0.20
 -0.85  -1.00   1.71
 -0.60   0.40   0.40
  0.40  -0.60   0.40
  0.40   0.40  -0.60
```

Now we can use matrix variable handling commands to define the variables *A* and *B*:

>> A = matrix (1:3, 1:3)

```
A =

  1.67  -0.12  -0.93
 -0.42   1.17   0.20
 -0.85  -1.00   1.71
```

>> B = matrix (4:6, 1:3)

```
B =

 -0.60  0.40  0.40
  0.40 -0.60  0.40
  0.40  0.40 -0.60
```

Commands that Work with Files in the Operational Environment

There is a group of commands that are used to work with files, allowing you to analyze, copy, delete, edit, and save data, among other options. These commands also allow the DOS environment to interrelate with the MATLAB environment, accommodating commands from both the operating system and from within the MATLAB Command Window.

Below is a list of these types of commands.

beep	*Produces a beep.*
CD directory	*Changes from the current directory to the given work directory.*
copy file f1 f2	*Copy the file (or directory) from the origin f1 to the destination file f2.*
delete file	*Delete the specified file (or graphic object).*
diary ('file')	*Writes the inputs and outputs of the current session in the file.*
dir	*Displays the files in the current directory.*
dos command	*Executes a DOS command and returns the result.*
edit M-file	*Edit an M-file.*
[path,name,ext,ver] = fileparts('file')	*Returns the path, name, extension and version of the specified file.*
file browser	*Displays the files in the current directory in a browser.*
fullfile('d1', 'd2',..., 'f')	*Builds a full file specification from the folders and file names specified.*
info toolbox	*Displays information about the specified toolbox.*
[M, X, J] =inmem	*Returns M-files, MEX-files and Java classes in memory.*
ls	*List the current directory in UNIX.*
MATLAB root	*Returns the name of the directory where MATLAB is installed.*
mkdir Directory	*Constructs a new directory.*
open('file')	*Opens the specified file.*
pwd	*Displays the current directory.*
tempdir	*Returns the name of the temporary directory of the system.*
name =tempname	*Assigns a unique name to the temporary directory.*
unix command	*Runs a UNIX command and returns the result.*
! command	*Executes an operating system command.*

Here are some examples:

```
>> dir
```

```
.              ..              matrix.ASC
```

```
>> ! dir
```

```
The volume of drive D has no label.
The volume serial number £ n is: 1179-07DC

Directory of D:\MATLABR12\work
```

```
01/01/2001 07:01 < DIR >.
2001-01-01 07:01 < DIR >..
02/01/2001 03:27 300 matrix.asc
1 files 300 bytes
2 dirs 1.338.146.816 bytes free
```

>> ! matrix.asc type
```
1. 6740445e + 000 - 1. 1964440e-001 - 9. 2759516e-001
-4 1647244e-001 1. 1737582e + 000 2. 0499870e-001
5035677e-001 - 8 - 1. 0006147e + 000 1. 7125190e + 000
-6 0000000e-001 4. 0000000e-001 4. 0000000e-001
4. 0000000e-001 - 6. 0000000e-001 4. 0000000e-001
4. 0000000e-001 4. 0000000e-001 - 6. 0000000e-001
```

>> tempdir

```
ans =
```

```
C:\DOCUME~1\CPL\CONFIG~1\Temp\
```

>> MATLABroot

```
ans =
```

```
D:\MATLABR12
```

>> pwd

```
ans =
```

```
D:\MATLABR12\work
```

>> cd ..
>> pwd

```
ans =
```

D:\MATLABR12

>> cd work
>> pwd

```
ans =
```

```
D:\MATLABR12\work
```

>> copyfile matrix.asc matrix1.asc
>> dir

```
.              ..              Matrix.ASC matrix1.asc
```
>> two dir

```
The volume of drive D has no label.
The volume serial number £ n is: 1179-07DC

Directory of D:\MATLABR12\work

01/01/2001 07:01 < DIR >.
01/01/2001 07:01 < DIR >...
02/01/2001 03:27 300 matrix.asc
02/01/2001 03:27 300 matrix1.asc
                2 files 600 bytes
2 dirs 1.338.130.432 bytes free
```

An important command that allows direct editing in a window of any M-file is *edit*. The figure below shows the edit window for the file *matrix1.asc*.

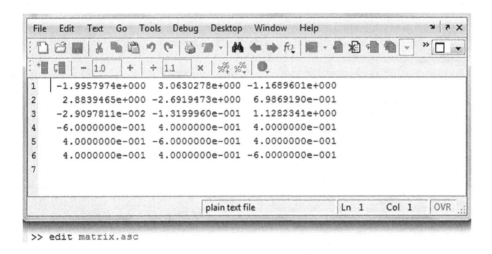

```
>> edit matrix.asc
```

Commands that Handle Functions

The list below describes a group of commands that handle functions, displaying help on them, providing access to information, and generating reports in MATLAB.

addpath('dir', 'dir2',...)	*Adds the directories to the MATLAB search path.*
doc **doc file** **doc toolbox/** **doc toolbox/function**	*Displays HTML documentation in the help panel for MATLAB functions in the Command Window, for a specified M-file, for the contents of a specified toolbox or for specified toolbox functions.*
help **help file** **help toolbox/** **help toolbox/function**	*Displays help for MATLAB functions in the Command Window, for a specified M-file, for the contents of a specified toolbox or for specified toolbox functions.*

(*continued*)

helpbrowser	*Shows the MATLAB help browser.*
helpdesk	*Shows the help browser located on the home page.*
helpwin	*Displays help for all MATLAB functions.*
docopt	*Shows the location of the UNIX help file.*
genpath	*Generates a path string.*
lasterr	*Returns the last error message.*
lastwarn	*Returns the last warning message.*
license	*Displays the MATLAB license number.*
lookfor theme	*Shows all functions related to search.*
partial pathname	*A partial pathname is a pathname relative to the MATLAB path matlabpath that is used to locate private and method files which are usually hidden or to restrict the search for files when more than one file with the given name exists.*
path	*Displays the complete path to MATLAB.*
pathtool	*Displays the complete path to MATLAB in windowed mode.*
profile	*Starts the profiler utility, to debug and optimize M-files code.*
profreport	*Generates a profile report in HTML format and suspends the windows profiler utility.*
rehash	*Refreshes caches of system files and functions.*
rmpath directory	*Removes the path from the MATLAB directory.*
support	*Opens the MathWorks website.*
typefile	*Lists the contents of the file.*
see (or see toolbox)	*Displays the version of MATLAB, Simulink and toolboxes.*
version	*Displays the version number of MATLAB.*
WebURL	*Directs the browser to the indicated Web address.*
what	*Lists MATLAB-specific files (.m, .mat, .mex .mdl and. p) in the current directory.*
whatsnew	*Shows help files with news of MATLAB and its toolboxes.*
which function	*Locates functions.*
which file	*Locates files.*

Here are some examples:

```
>> version

ans =

6.1.0.450 (R12.1)
```

```
>> license

ans =

DEMO
```

>> help toolbox\symbolic

```
Symbolic Math Toolbox.
Version 2.1.2 (R12.1) 11-Sep-2000

New Features.
Readme     - Overview of the new features in/changes made to
the Symbolic and Extended Symbolic Math Toolboxes.

Calculus.
diff       - Differentiate.
int        - Integrate.
limit      - Limit.
taylor     - Taylor series.
jacobian   - Jacobian matrix.
symsum     - Summation of series.

Linear Algebra.
diag       - Create or extract diagonals.
triu       - Upper triangle.
tril       - Lower triangle.
inv        - Matrix inverse.
det        - Determinant.
rank       - Rank.
rref       - Reduced row echelon form.
null       - Basis for null space.
colspace   - Basis for column space.
eig        - Eigenvalues and eigenvectors.
svd        - Singular values and singular vectors.
Jordan     - Jordan canonical (standard) form.
poly       - Characteristic polynomial.
expm       - Matrix exponential.
```

>> help int

```
--- help for sym/int.m ---

INT Integrate.
INT(S) is the indefinite integral of S with respect to its symbolic
variable as defined by FINDSYM. S is a SYM (matrix or scalar).
If S is a constant, the integral is with respect to 'x'.
INT(S,v) is the indefinite integral of S with respect to v. v is a
scalar SYM.
INT(S,a,b) is the definite integral of S with respect to its
symbolic variable from a to b. a and b are each double or
```

symbolic scalars.
INT(S,v,a,b) is the definite integral of S with respect to v
from a to b.

Examples:
syms x alpha u t;
int(1/(1+x^2)) returns atan(x)
int (sin(alpha*u), alpha) returns - cos(alpha*u) /u
int (4 * x * t, x, 2, sin (t)) returns 2 * sin (t) ^ 2 * t - 8 * t

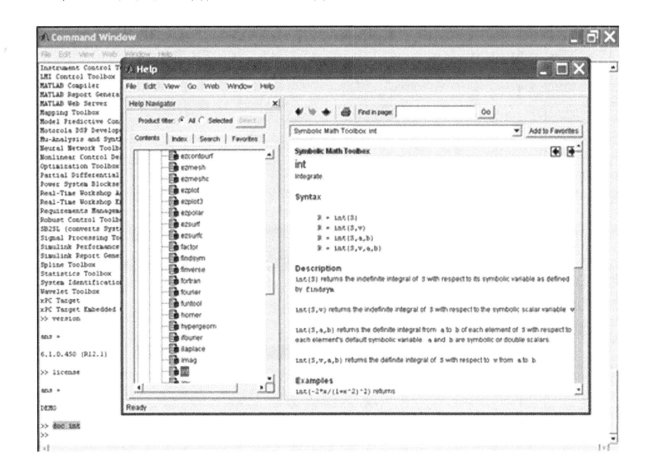

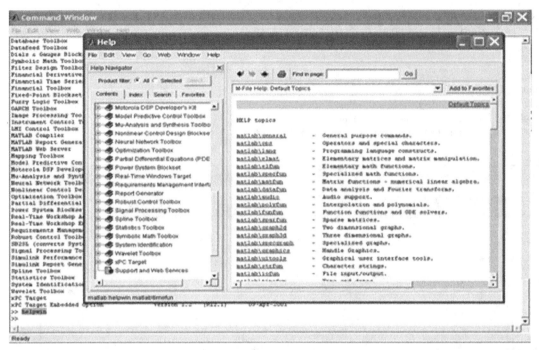

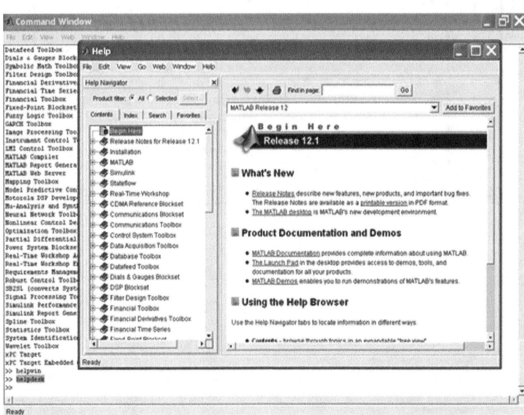

>> lookfor GALOIS

GFADD Add polynomials over a Galois field.
GFCONV Multiply polynomials over a Galois field.
GFCOSETS Produce cyclotomic cosets for a Galois field.
GFDECONV Divide polynomials over a Galois field.
GFDIV Divide elements of a Galois field.
GFFILTER Filter data using polynomials over a prime Galois field.
GFLINEQ Find a particular solution of Ax = b over a prime Galois field.
GFMINPOL Find the minimal polynomial of an element of a Galois field.
GFMUL Multiply elements of a Galois field.
GFPLUS Add elements of a Galois field of characteristic two.
GFPRIMCK Check whether a polynomial over a Galois field is primitive.
GFPRIMDF Provide default primitive polynomials for a Galois field.
GFPRIMFD Find primitive polynomials for a Galois field.
GFRANK Compute the rank of a matrix over a Galois field.
GFROOTS Find roots of a polynomial over a prime Galois field.
GFSUB Subtract polynomials over a Galois field.
GFTUPLE Simplify or convert the format of elements of a Galois field.

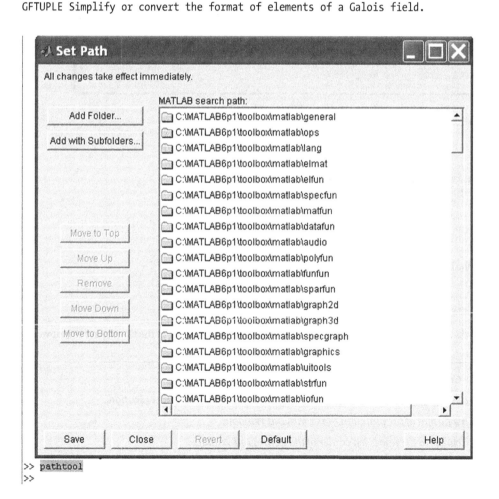

```
>> pathtool
>>
```

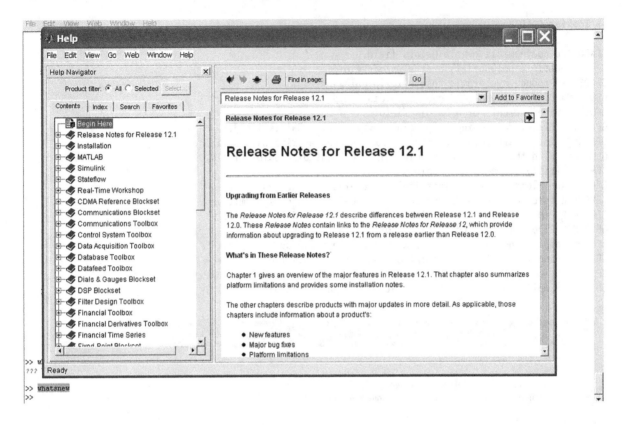

```
>> what

M-files in the current directory C:\MATLAB6p1\work

cosint
```

>> which sinint

```
C:\MATLAB6p1\toolbox\symbolic\sinint.m
```

Commands that Control the Command Window

The following table summarizes a group of commands in MATLAB which control the output in the Command Window.

CLC	*Clears the Command Window.*
echo	*Displays (echo on) or hides (echo off) the lines of an M-file code during its execution.*
format type	*Controls the format of the output in the Command Window.*
home	*Moves the cursor to the upper left corner of the Command Window.*
more	*Enables paging of the output in the Command Window.*

The possible types for the *format* command are given below:

Type	Result	Example
+	*+,-, white*	+
bank	*Fixed to dollars and cents.*	3.14
compact	*Suppresses excess line feeds in the output. Contrast this with loose.*	Theta = pi /2
Hex	*Hexadecimal format.*	400921fb54442d18
long	*15 digit fixed-point.*	3.14159265358979
long e	*15 digit floating-point.*	3.141592653589793e + 00
long g	*15 significant digits (fixed or floating point).*	3.14159265358979
loose	*Adds line feeds to make the output more readable. Contrast this with compact.*	Theta = 1.5708
rat	*Rational format.*	355/113
short	*5 digit fixed-point.*	3.1416
short e	*5 digit floating-point.*	3. 1416e + 00
short g	*5 significant digits (fixed or floating-point)*	3.1416

Start and Exit Commands

MATLAB offers the following start and exit commands.

finish	*Complete an M-file.*
exit	*Finish MATLAB.*
MATLAB	*Start MATLAB (only on UNIX).*
MATLABrc	*Start an M-file.*
quit	*Finish MATLAB.*
startup	*Start an M-file.*

File Input/Output Commands

MATLAB has a group of so-called input/output commands which operate on files, allowing the user to open and close files, read and write to files, control the position in a file and export and import data. The following table summarizes these commands. Their full syntax will be described in the following paragraphs.

Opening and closing files

fclose	*Closes one or more files.*
fopen	*Opens a file or obtains information about open files.*

Plain input/output

fread	*Reads binary data from a file.*
fwrite	*Writes binary data to a file.*

Format input /output

fgetl	*Returns the next line of a file as a string without ends of lines.*
fgets	*Returns the next line of a file as a string with ends of lines.*
fprintf	*Types formatted data into a file.*
fscanf	*Reads formatted data from a file.*

Controlling position in a file

feof	*Tests for the end of file.*
ferror	*Returns the error message for the most recent input/output operation on a specified file.*
frewind	*Rereads an open file.*
fseek	*Moves the location of a file position indicator.*
ftell	*Finds the location of a file position indicator.*

String conversion

sprintf	*Type data formatted as a string.*
sscanf	*Read under the control of format strings.*

Specialized input/output functions

dlmread	*Reads files with delimited ASCII format.*
dlmwrite	*Writes files with delimited ASCII format.*
hdf	*HDF interface.*
imfinfo	*Returns information about graphics files.*
imread	*Reads images from graphics files.*
imwrite	*Writes an image in a graphics file.*
strread	*Reads formatted data from a string.*
textread	*Reads formatted data from a text file.*
wk1read	*Reads data from Lotus123 WK1 spreadsheet files.*
wk1write	*Writes data in Lotus123 WK1 worksheet files.*

Opening and Closing Files

In order to read or write data to a file (which does not have to be in ASCII or MATLAB format), first use the command *fopen* to open it. Then, to perform read or write operations on it, use the corresponding read and write commands (*fload, fwrite, fprintf, import* etc.). Finally, use the command *fclose* to close the file. The file that is opened may be new or may be an existing file which is to be accessed either to broaden its content or simply to read it.

The command *fopen* returns a file that consists of a non-negative integer which is assigned by the operating system to the opened file. This file identifier is used as a reference for the subsequent management of the open file as it is read (*read*), written to (*write*) or closed (*close*). If the file does not open correctly, *fopen* returns - 1 as the file identifier. As a generic file identifier, *fidelity* is commonly used. The syntax of the commands *fopen* and *fclose* is described below.

fid = fopen ('file')	*Opens the specified existing file.*
fid = fopen ('file', 'permission')	*Opens the file for the given permission type.*
[fid, message] = fopen('file', 'permission', 'architecture')	*Opens the file for the given permission and with the numerical format of the architecture.*
fids = fopen ('all')	*Returns a column vector with the identifiers of all open files*
[filename, permission, architecture] = fopen(fid)	*Returns the name of the file, the type of permission and the numerical format of the specified architecture relating to the file whose ID is fid.*
fclose (fid)	*Closes the identifier fid file if it is open. Returns 0 if the process has been performed successfully and -1 otherwise.*
fclose ('all')	*Closes all open files. Returns 0 if the process has been performed successfully and -1 otherwise.*

The possible types of permissions are the following:

'r'	*Open the existing file for reading (this is the default permission).*
'r +'	*Open the existing file for reading and writing.*
'w'	*Creates the new file and opens it for writing, and if there is already a file with that name, deletes it and opens it again as an empty file.*
'w +'	*Creates the new file for reading and writing, and if there is already a file with that name, deletes it and opens it again as an empty file.*
'a'	*Creates the new file and opens it for writing, and if there is already a file with that name, adds new content at the end of the existing file.*
'a +'	*Creates the new file and opens it for reading and writing, and if there is already a file with that name, adds new content to the end of the existing file.*
'A'	*Append without automatic flushing of the current output buffer. (Used with tape drives.)*
'W'	*Write without automatic flushing of the current output buffer. (Used with tape drives.)*

Possible architectures for the numerical format types are as follows:

'native' or 'n'	*Numeric format of the current machine.*
'ieee-le' or 'l'	*Small-format IEEE floating-point.*
'ieee-be' or 'b'	*Large format IEEE floating-point.*
'vaxd' or 'd'	*VAX D floating-point format.*
'vaxg' or 'g'	*VAX G floating-point format.*
'cray' or 'c'	*Large type Cray floating-point format.*
'ieee-le.164' or 'a'	*Small format IEEE floating-point and 64-bit data length.*
'ieee-be. l64' or 's'	*IEEE floating-point, 64-bit data length large format.*

Being able to open a file according to the numerical format of a given architecture allows it to be used in different MATLAB platforms.

Reading and Writing Binary Files

Reading and writing binary files is done via the commands *fwrite* and *fread*. The command *fwrite* is used to write binary data to a file previously opened with the command *fopen*. The command *fread* is used to read data from a binary file previously opened with the command *fopen*. Its syntax is as follows:

fwrite (fid, A, precision)	*Writes the specified items in A (which in general is an array) in the file identifier fid (previously opened) with the specified accuracy.*
A = fread (fid)	*Reads the data from the binary file opened with identifier fid and writes them to the matrix A, which by default will be a column vector.*
[A, count] = fread(fid, size, precision)	*Reads the data from the file identifier fid with the dimension specified in size and precision given by* precision, *and writes them to a matrix A of dimension* size *and whose total number of elements is* count.

The specification *size* is optional. If *size* is set to n, *fread* reads the first n data from the file (by columns and in order) as a column vector, A, of length n. If *size* is set to *inf*, *fread* reads all file data by columns and in order, to form a single column vector A (this is the default value). If *size* is set to $[m, n]$, *fread* reads $m \times n$ file elements by columns and in order, completing the matrix A of dimension $(m \times n)$. If there are insufficient elements in the file to complete the matrix, it will be completed with zeros.

The argument *precision* is relative to the numeric precision of the machine on which you are working and may present different values. In addition to its own types of formatting for numerical precision, MATLAB also accepts those of the programming languages C and FORTRAN. Below is a table with the possible values of *precision*.

MATLAB	C or FORTRAN	Interpretation
'schar'	'signed char'	*Character with sign; 8-bit*
'uchar'	'unsigned char'	*Character unsigned; 8-bit*
'int8'	'integer * 1'	*Integer; 8-bit*
'int16'	'integer * 2'	*Integer; 16-bit*
'int32'	'integer * 4'	*Integer; 32-bit*
'int64'	'integer * 8'	*Integer; 64-bit*
'uint8'	'integer * 1'	*Unsigned integer; 8-bit*
'uint16'	'integer * 2'	*Unsigned integer; 16-bit*
'uint32'	'integer * 4'	*Unsigned integer; 32-bit*
'uint64'	'integer * 8'	*Unsigned integer; 64-bit*
'float32'	'real * 4'	*Floating point; 32-bit*
'float64'	'real * 8'	*Floating point; 64-bit*
'double'	'real * 8'	*Floating point; 64-bit*

The following formats are also supported by MATLAB, but there is no guarantee that the same size will be used on all platforms.

MATLAB	C or FORTRAN	Interpretation
'char'	'char * 1'	*Character; 8-bit*
'short'	'short'	*Integer; 16-bit*
'int'	'int'	*Integer; 32-bit*
'long'	'long'	*Integer; 32 or 64 bit*
'ushort'	'unsigned short'	*Unsigned integer; 16-bit*
'uint'	'unsigned int'	*Unsigned integer; 32-bit*
'ulong'	'unsigned long'	*Unsigned integer; 32 or 64 bit*
'float'	'float'	*Floating point; 32-bit*
'intN'		*Whole width N integer bits ($1 \leq N \leq 64$)*
'ubitN'		*Integer unsigned width N bits ($1 \leq N \leq 64$)*

When they are read and stored, formats often use the implication symbol as illustrated in the following examples:

' **uint8 = > uint8'**	*Reads entire 8-bit unsigned integers and stores them in an array of unsigned 8-bit integers.*
' *** uint8'**	*An abridged version of the previous example.*
' **bit4 = > int8'**	*Reads entire 4 bit signed integers packaged in bytes and stores them in an array of 8-bit integers. Each 4-bit integer is converted to an 8-bit integer.*
' **double = > real * 4'**	*Reads double precision floating point numbers and stores them in an array of 32-bit real floating point numbers.*

As a first example we can view the contents of the file *fclose.m* using the command *type* as follows:

```
>> type fclose.m

%FCLOSE Close file.
%   ST = FCLOSE(FID) closes the file with file identifier FID,
%   which is an integer obtained from an earlier FOPEN.  FCLOSE
%   returns 0 if successful and -1 if not.
%
%   ST = FCLOSE('all') closes all open files, except 0, 1 and 2.
%
%   See also FOPEN, FREWIND, FREAD, FWRITE.
%   Copyright 1984-2001 The MathWorks, Inc.
%   $Revision: 5.8 $  $Date: 2001/04/15 12:02:12 $
% Built-in function.
```

This is equivalent to using the command *type* before opening the file with *fopen,* followed by reading its contents with *fread* and presenting it with the function *char.*

```
>> fid = fopen('fclose.m','r');
>> F = fread(fid);
>> s = char(F')

s =

%FCLOSE Close file.
%   ST = FCLOSE(FID) closes the file with file identifier FID,
%   which is an integer obtained from an earlier FOPEN. FCLOSE
%   returns 0 if successful and -1 if not.
%
%   ST = FCLOSE('all') closes all open files, except 0, 1 and 2.
%
%   See also FOPEN, FREWIND, FREAD, FWRITE.
%   Copyright 1984-2001 The MathWorks, Inc.
%   $Revision: 5.8 $  $Date: 2001/04/15 12:02:12 $
% Built-in function.
```

In the following example, we create a binary file *id4.bin* which contains the 16 elements of the identity matrix of order 4 stored in 4 byte integers (64 bytes in total). First we open the file which will contain the matrix, with permission to read and write, and then write the matrix to the file with the appropriate format. Finally, we close the open file.

```
>> fid = fopen ('id4. bin ',' w +')

fid =

5

>> fwrite(fid,eye(4),'integer*4')

ans =

16

>> fclose (5)

ans =

0
```

In the previous example, when the file was opened, the system assigned ID 5 to it. After writing the matrix to the file, it was necessary to close it with the command *fclose* using the indicator. The answer of zero means the closure has been successful.

If we now want to see the contents of the binary file just recorded, we open it, with reading permission, by using the command *fopen* and read its elements with *fread*, in the same matrix structure and format in which it was saved.

```
>> fid = fopen('id4.bin','r+')

fid =

5

>> [R,count]=fread(5,[4,4],'integer*4')

R =
1 0 0 0
0 1 0 0
0 0 1 0
0 0 0 1

count =

16
```

Reading and Writing Formatted ASCII Text Files

It is possible to write formatted text to a file previously opened with the command *fopen* (or to the screen itself) using the command *fprintf*. On the other hand, it is possible, using the command *import*, to read formatted data from a file previously opened with the command *fopen*. The syntax is as follows:

fprintf(fid, 'format', A,...)	*Writes the specified items in A (which in general is an array) in the file identifier fid (previously opened) with the format specified in 'format'.*
fprintf('format', A,...)	*Writes to the screen.*
[A, count] = fscanf(fid, 'format')	*Reads the data in the given format of an open file with identifier fid and writes them to the matrix A, which by default will be a column vector.*
[A, count] = fscanf(fid, 'format', size)	*Reads the data from the file identifier fid with the specified size and format, and writes them to a matrix A of dimension size and whose number of elements is count.*

The argument *format* consists of a chain (preceded by the character '\') formed by characters and conversion characters according to the different formats (preceded by the character '%').

The possible characters are as follows:

\n	*Executes the step to a new line.*
\t	*Executes a horizontal tab.*
\b	*Executes a step backward from a single character (backspace), deleting the current character.*
\r	*Executes a carriage return.*
\f	*Executes a page jump (form feed).*
****	*Executes a backslash.*
\'	*Executes a single quotation mark.*

Possible conversion characters are the following:

%d	*Decimal integers*
%o	*Octal integers*
%x	*Hexadecimal integers*
%u	*Unsigned decimal integers*
%f	*Real fixed-point*
%e	*Real floating-point*
%g	*Use whichever of d, e or f has the greater precision in the minimum of space*
%c	*Individual characters*
%s	*Character string*
%E	*Real floating point (uppercase E)*
%X	*Uppercase hexadecimal notation*
%G	*%g format with capital letters*

When working with integers, conversion characters are used in the form % *nv* (*n* is the number of digits of the integer and *v* is the conversion character, which can be *d*, *o*, *x* or *u*). For example, the format % *7x* indicates a hexadecimal integer with 7 digits.

When working with real numbers, conversion characters are used in the form %*n.mv* (*n* is the total number of digits of the real number including the decimal point, *m* is the number of decimal places of the real number and *v* is the conversion character, which can be *f*, *e* or *g*). For example, the format %*6.2f* indicates a fixed point real number having 6 numbers in total (including the point) and with 2 decimal places.

When working with strings, conversion characters are used in the form % *na* (*n* is the total number of characters in the string and *a* is the conversion character, which can be *c* or *s*). For example, the format % *8s* indicates a string of 8 characters.

In addition, escape characters and conversion of the C language are supported (see C manuals for further information).

In the *import* command the *size* preference is optional. If *size* is set to *n*, *import* reads the first *n* data from the file (by columns and in order) as a vector column *A* of length *n*. If *size* is set to *inf*, *fread* reads all file data by columns and in order, to form a single column vector *A* (this is the default value). If *size* is set to *[m, n]*, *fread* reads *m×n* file elements by columns and in order, completing the matrix *A* of dimension *(m×n)*. If there are insufficient elements in the file, the matrix is completed with zeros as needed. The argument *format* takes the same values as the command *fprintf*.

For reading ASCII files there are two other commands, *fgetl* and *fgets*, which present different lines of a text file as a string. Its syntax is as follows:

fgetl (fid)	*Reads the characters in the text with file identifier fid line by line, ignoring carriage returns, and returns them as a string.*
fgets (fid)	*Reads the characters in the text with file identifier fid line by line, including carriage returns, and returns them as a string.*
fgets (fid, nchar)	*Returns at least nchar characters in the next line.*

As an example we create an ASCII file *exponen.txt,* which contains the values of the exponential function for values of the variable between 0 and 1 separated by 0.1.

The format of the text in the file should consist of two columns of real floating point numbers, in such a way that the values of the variable appear in the first column and the corresponding values of the exponential function appear in the second column. Finally, we issue commands to display the contents of the file on screen.

```
>> x = 0:.1:1;
>> y= [x; exp(x)];
>> fid=fopen('exponen.txt','w');
>> fprintf(fid,'%6.2f  %12.8f\n', y);
>> fclose(fid)

ans =

0
```

Now information is presented directly on screen without having to save it to disk:

```
>> x = 0:. 1:1;
>> y = [x; exp (x)];
(>> fprintf('%6.2f. 8f\n', and)12%
```

```
0.00 1.00000000
0.10 1.10517092
0.20 1.22140276
0.30 1.34985881
0.40 1.49182470
0.50 1.64872127
0.60 1.82211880
0.70 2.01375271
0.80 2.22554093
0.90 2.45960311
1.00 2.71828183
```

We then read the newly generated ASCII file *exponen.txt,* so that the format of the text must consist of two columns of real numbers with maximum precision in the minimum of space, the first column showing the values of the variable and the second showing the corresponding values of the exponential function.

```
>> fid=fopen('exponen.txt');
>> a = fscanf(fid,'%g  %g', [2 inf]);
>> a = a '
```

```
a =
```

```
0 1.0000
0.1000 1.1052
0.2000 1.2214
0.3000 1.3499
0.4000 1.4918
0.5000 1.6487
0.6000 1.8221
0.7000 2.0138
0.8000 2.2255
0.9000 2.4596
1.0000 2.7183
```

We then open the file *exponent.txt* and read its contents line by line with the command *fgetl.*

```
>> fid=fopen('exponen.txt');
>> linea1=fgetl(fid)
```

```
linea1 =
```

```
0.00    1.00000000
```

```
>> linea2=fgetl(fid)
```

```
linea2 =
```

```
0.10 1.10517092
```

Below, the command *sprintf* outputs a string variable that presents the given text according to the specified format together with the value of the golden ratio.

```
>> S = sprintf ('the golden ratio is % 6.3f,' (1 + sqrt (5)) / 2).

S =

the golden ratio is 1.618
```

Finally we generate a column vector whose two elements are approximations of the irrational numbers e and π.

```
>> S = '2.7183 3.1416';
>> A = sscanf(S,'%f')

A =

2.7183
3.1416
```

Control Over the File Position

The commands *fseek, ftell, feof, frewind* and *ferror* control position in the file. The command *fseek* allows you to move the position indicator in a previously opened file. The command *ftell* returns the current status of the position indicator within a file. The command *feof* indicates whether the position indicator is located at the end of the file. The command *frewind* places the position indicator at the beginning of the file. The command *arenas* returns the error message associated with the most recent input or output operation on a specified file previously opened with *fopen*. The syntax of these commands is as follows:

fseek(fid, n, 'origin')	*Moves the position indicator n bytes from the source indicated by the argument origin within the file identifier fid previously opened with fopen. If n > 0, the position indicator moves n bytes forward towards the end of the file. If n < 0, the position indicator moves n bytes backward towards the beginning of the file. If n = 0, the position indicator does not change. The values that the argument origin can take are: 'bof' or - 1 (the origin is at the beginning of the file), 'cof' or 0 (the source is at the current position of the indicator) and 'eof' or 1 (the source is at the end of the file).*
n = ftell (fid)	*Returns the number of bytes from the beginning of the file whose identifier is fid (previously opened with fopen) to the current position indicator.*
feof (fid)	*Returns 1 if the position indicator is located at the end of the file with identifier fid (previously opened) and 0 otherwise.*
frewind (fid)	*Places the position indicator at the beginning of the (previously opened) file with identifier fid.*
ferror (fid) output	*Returns the (possibly empty) error message associated with the most recent input or output operation on the previously opened file with identifier fid.*
[message, errnum] =ferror (fid)	*In addition to the error message, this returns its error number. An error number of 0 indicates that the error message is empty, i.e. the most recent input or output operation did not result in an error.*

As an example, we write the two-byte integers from 1 to 5 into a binary file named *five.bin*. We check the status of the position indicator in the file and move 6 bytes forward, checking that the operation has been correctly carried out. Subsequently we will move the position indicator 4 bytes backwards and find which number has been located.

```
>> A=[1:5];
fid=fopen('five.bin','w');
fwrite(fid,A,'short');
fclose(fid);
fid=fopen('five.bin','r');
n = ftell (fid)

n =

0
```

As the number of bytes from the beginning of the file to the current location of the position indicator is revealed to be $n = 0$, the position indicator is obviously located at the beginning of the file, i.e. at the first value, which is 1. Another way to see that the position indicator is located on 1 is to use the command *fread* to read only the first element of the binary file *five.bin*:

```
>> fid=fopen('five.bin','r');
principal = fread(fid,1,'short')

principal =

1
```

Now we are going to move the position indicator 6 bytes forward and check the new position:

```
>> fid=fopen('five.bin','r');
fseek(fid,6,'bof');
n=ftell(fid)

n =

6
```

```
>> principal=fread(fid,1,'short')

principal =

4
```

We have seen that the position indicator has moved 6 bytes to the right, landing on the element 4 (bear in mind that each file element occupies 2 bytes). Now we are going to move the position indicator 4 units to the left and determine on which item it has been moved to:

```
>> fseek(fid,-4,'cof');
n=ftell(fid)

n =

4
```

```
>> principal=fread(fid,1,'short')
```

principal =

3

Finally, the position indicator has been set to 4 bytes from the beginning of the file, i.e. on element 3 (again recalling that each file element occupies 2 bytes).

Exporting and Importing Data to Lotus 123 and Delimited ASCII String and Graphic Formats

There is a group of commands in MATLAB which enable you to export and import data between Lotus 123 and MATLAB. Another group of commands allows you to export and import data between ASCII files with delimiters and MATLAB. The following table summarizes these commands.

A = wk1read (file)	*Reads the Lotus 123 spreadsheet named file.wk1y and imports it as a MATLAB matrix whose rows and columns are those of the worksheet.*
A = wk1read(file, F,C)	*Reads the Lotus 123 spreadsheet named file.wk1 from row F and column C, and imports it as a MATLAB matrix whose rows and columns are those of the worksheet.*
A = wk1read(file, F,C,R)	*Reads the R data range of the Lotus 123 spreadsheet named file.wk1 from row F and column C, and imports it as a MATLAB matrix whose rows and columns are those of the worksheet.*
A = wk1write (file, M)	*Enters the MATLAB matrix M as a Lotus 123 spreadsheet file named file.wk1 whose rows and columns are those of the matrix M.*
A = wk1write(file, M,F,C)	*Enters the MATLAB matrix M as a Lotus 123 spreadsheet file named file.wk1 whose rows and columns are those of the matrix M starting at row F and column C.*
M = dlmread (file, D)	*Reads the specified formatted file whose data are separated by the delimiter D and returns it as the matrix M.*
M = dlmread(file, D,F,C)	*Reads the specified files whose data are separated by the delimiter D and returns it as the matrix M which begins at F row and column C.*
M = dlmwrite (file, M,D)	*Writes the matrix M in the specified formatted file, whose data are separated by the delimiter D.*
M = dlmwrite(file, D,F,C)	*Writes the matrix M, starting at row F and column C, in the specified formatted file, whose data are separated by the delimiter D.*

(*continued*)

A = **imread**(file,*fmt*)	*Reads the image in a graphical format fmt file given in grayscale or true color.*
[X,map] = **imread**(file,*fmt*)	*Reads the image in graphical format fmt of the given file indexed in X and its associated map colors.*
[...] = **imread** (file)	*Tries to infer the format of the file from its content.*
[...] = **imread**(...,idx) **(CUR, ICO and TIFF only)**	*Reads an image of order idx in a TIFF, CUR or ICO file.*
[...] = **imread**(...,idx) **(HDF only)**	*Reads an image of order idx in an HDF file.*
	Reads an image with background color and intensity of a given grayscale.
[...] = **imread**(...,'backgroundcolor', BG) **(PNG only)**	*Reads an image in graphical format from the given file fmt applying transparency mask.*
[A,map,alpha] = **imread(file, fmt...)**	*Returns the transparency mask.*
[map, alpha] = **imread** (...) **(PNG only)**	
imwrite(A, file, *fmt*) **imwrite (X, map, file, *fmt*)**	*Writes the image in graphical format fmt in the given file in grayscale or true color.*
imwrite(...,filename) **imwrite(...,param1,val1,** **param2, val2...)**	*Writes the indexed image in X and its associated color map in the given file in graphic format fmt.*
	Writes the image in the given file, inferring the format of filename from its extension.
	Specifies the control of various characteristics of the output file parameters.
info = **imfinfo**(file,*fmt*)	*Provides information on the graphic file format fmt.*
A = **strread**('C')	*Reads the C string numeric data.*
A = **strread**('C','',N)	*Reads N lines of the C string numeric data.*
A = **strread**('C','',p,value,...)	*Reads the C string data according to the parameter p and value.*
A = **strread**('str','',N,p,value,...)	*Reads N rows of C according to the parameter p and value.*
[A,B,C,...]= **strread**('C','format')	*Reads the string C with the specified format.*
[A,B,C,...] = **strread** ('C','format',N)	*Reads N lines of the string C with the specified format.*
[A,B,C,...] = **strread** ('C','format',p,value,...)	*Reads the C string with the specified format according to the parameter p and value.*
[A,B,C,...] = **strread** ('C','format',N,param,value,...)	*Reads N lines of the C string with the specified format according to the parameter p and value.*
[A,B,C,...] = **textread**('file','format')	*Reads data from the text file using the given format.*
[A,B,C,...] = **textread**('file','format',N)	*Reads data from the text file using the given format N times.*
[...] = **textread**(...,'p','value',...)	*Reads measurement data using the specified parameter and value.*

Possible values for the *fmt* file graphic format are presented in the following table:

Format	Type of file
'bmp'	*Windows Bitmap (BMP)*
'cur'	*Windows Cursor (CUR) resources*
'hdf'	*Hierarchical Data Format (HDF)*
'ico'	*Windows Icon (ICO) resources*
'jpg' or 'jpeg'	*Joint Photographic Experts Group (JPEG)*
'pcx'	*Windows Paintbrush (PCX)*
'png'	*Portable Network Graphics (PNG)*
'tif' or 'tiff'	*Tagged Image File Format (TIFF)*
'xwd'	*X Windows Dump (XWD)*

The following table shows the types of image that *imread* can read.

Format	Variants
BMP	*1- bit, 4-bit, 8-bit, 24 - bit images without compression; 4-bit images with compression (RLE) 8 - bit*
CUR	*1- bit, 4-bit and 8-bit images without compression*
HDF	*8- bit with or without associated color map image data sets; 24-bit and 8-bit data image sets*
ICO	*1- bit, 4-bit and 8-bit images without compression*
JPEG	*Any baseline JPEG image (8 or 24-bit); JPEG images with any commonly used extension*
PCX	*1-bit, 8-bit, and 24-bit images*
PNG	*Any PNG image, including 1-bit, 2-bit, 4-bit, 8-bit, and 16-bit images in grey scales; 8-bit and 16-bit indexed images; 24-bit and 48-bit RGB images*
TIFF	*Any baseline TIFF image, including 1-bit 8-bit and 24-bit images without compression; 1-bit, 8-bit, 16-bit and 24-bit compressed images; 1-bit images compressed with CCITT; also 16-bit greyscale, 16-bit indexed and 48-bit RGB images*
XWD	*1-bit and 8-bit ZPixmaps; XYBitmaps; 1-bit XYPixmaps*

The following table shows all the formats that support the commands *strread* and *testread*.

Format	Action	Output
Literals (characters)	*Ignores correspondence characters*	*No*
%d	*Reads a signed integer value*	*Double array*
%u	*Reads an integer value*	*Double array*
%f	*Reads a floating point value*	*Double array*
%s	*Reads with white space separation*	*Cell array of strings*

(continued)

Format	Action	Output
%q	Reads a string enclosed in double quotes	Cell array of strings. Excluding double quotes.
%c	Reads characters including blanks	Array character
%[...]	Reads the longer string containing the characters specified within square brackets	Cell array of strings
%[^...]	Reads the longer non-empty string containing characters not specified within square brackets	Cell array of strings
%* ... in place of %	Ignores the correspondence between characters specified by *	Without output
%w ... in place of %	Reads the specified field width w. The format %f supports % w.pf, where w is the width of the field and p is the precision.	

The possible pairs (parameter, value) that can be used as custom options for the *strread* and *testread* commands are presented in the following table:

Parameter	Value	Action
whitespace	Any of the following list	Characters, *, as white space. The default is \b\r\n\t.
	b	Backspace
	f	Form of the identifier
	n	New line
	r	Carriage return
	t	Horizontal tab
	\\	Backslash (moves backwards one space)
	\" or "	Mark with single quotes
	%%	Percent sign
delimiter	Delimiter character	Specifies the delimiter character
expchars	Character exponent	By default this is eEdD
bufsize	Positive integer	Maximum length of string in bytes (4095)
headerlines	Positive integer	Ignores the specified number of lines at the beginning of the file
Commentstyle	MATLAB	Ignore characters after %
Commentstyle	Shell	Ignore characters after #
Commentstyle	c	Ignored characters between / * and * /
Commentstyle	c ++	Ignore characters after / /

As a first example we read information from the file *canoe.tif*.

```
>> info = imfinfo ('canoe. tif')
```

```
Info =
Filename: 'C:\MATLAB6p1\toolbox\images\imdemos\canoe.tif'
FileModDate: '25-Oct-1996 23:10:40'
FileSize: 69708
Format: 'tif'
```

```
FormatVersion: []
Width: 346
Height: 207
BitDepth: 8
ColorType: 'indexed'
FormatSignature: [73 73 42 0]
ByteOrder: 'little-endian'
NewSubfileType: 0
BitsPerSample: 8
Compression: 'PackBits'
PhotometricInterpretation: 'RGB Palette'
StripOffsets: [9x1 double]
SamplesPerPixel: 1
RowsPerStrip: 23
StripByteCounts: [9x1 double]
XResolution: 72
YResolution: 72
ResolutionUnit: 'Inch'
Colormap: [256x3 double]
PlanarConfiguration: 'Chunky'
TileWidth: []
TileLength: []
TileOffsets: []
TileByteCounts: []
Orientation: 1
FillOrder: 1
GrayResponseUnit: 0.0100
MaxSampleValue: 255
MinSampleValue: 0
Thresholding: 1
```

The following example reads the sixth image of the file *flowers.tif*.

```
>> [X,map] = imread('flowers.tif',6);
```

The following example reads the fourth image of an HDF file.

```
>> info = imfinfo ('skull. hdf');
[X, map] = imread ('skull hdf',. info (4)Reference);
```

The following example reads a PNG image in 24-bit with complete transparency.

```
>> bg = [255 0 0];
A = imread('image.png','BackgroundColor',bg);
```

Below is an example with *sprintf* and *strread*.

```
>> s = sprintf('a,1,2\nb,3,4\n');
[a,b,c] = strread(s,'%s%d%d','delimiter',',')
```

```
a =

'a'
'b'

b =

1
3

c =

2
4
```

If the file *mydata.dat* has as first line *Sally Type1 12.34 45 Yes,* then the first column will be read in free format.

```
>> [names,types,x,y,answer] = textread('mydata.dat','%s %s %f ...
%d %s',1)

names =
        'Sally'
types =
        'Type1'
x =
        12.34000000000000
y =
        45
answer =
        'Yes'
```

We then use the command *strread.*

```
>> s = sprintf('a,1,2\nb,3,4\n');
[a,b,c] = strread(s,'%s%d%d','delimiter',',')

a =

'a'
'b'

b =

1
3

c =

2
4
```

Sound Processing Functions

MATLAB's Basic module includes a group of functions that read and write audio files. These functions are presented in the following table:

General sound functions	
μ=lin2mu(y)	*Converts a linear audio signal of amplitude - 1≤y≤1 to a μ-encoded audio signal with 0≤μ≤255.*
Y=mu2lin(μ)	*Converts a μ-encoded audio signal (μ≤255) to a linear audio signal (-1≤y≤1).*
sound(y,Fs)	*Converts the audio signal y to a sound at sample rate Fs.*
sound(y)	*Converts the audio signal y to a sound at the standard 8192 Hz sampling rate.*
sound(y,Fs,b)	*Using b bits/sample when converting the audio signal y to a sound at sample rate Fs.*
Workstations SPARC-specific functions	
auread('f.au')	*Reads the NeXT/SUN sound files f.au.*
[y,Fs,bits] = uread('f.au')	*Gives the sample rate in Hz and the number of bits per sample used to encrypt the data in the file f.au.*
auwrite (y, 'f.au')	*Writes a NeXT/SUN sound file f.au.*
auwrite(y, Fs, 'f.au')	*Writes a type f.au sound file and specifies the sample rate in Hertz.*
Functions of sound.WAV	
wavplay(y,Fs)	*Reproduces the audio signal y with sampling rate Fs.*
wavread('f.wav')	*Reads the f.wav sound files.*
[y,Fs,bits] = wavread('f.wav')	*Returns the sampling rate Fs and the number of bits per sample to read the f.wav sound file.*
wavrecord(n, Fs)	*Records samples of a digital audio signal at the sample rate n Fs.*
wavwrite(y,'f.wav')	*Writes a type f.wab sound file.*
wavwrite(y,Fs, 'f.wav')	*Writes a sound file f.wab with sampling rate Fs.*

EXERCISE 3-1

Construct a magic square of order 4, and write its inverse matrix in a binary file named magic.bin.

We start by defining the matrix:

```
>> M = magic (4)

M =

16 2 3 13
5 11 10 8
9 7 6 12
4 14 15 1
```

Then we open a file named *magic.bin*, with read/write permission to store the matrix *M*. We use the permission 'w +' because we want to open a new file, i.e. it does not already exist, and in addition we need to write to it (since the file does not already exist, we could also use the permission 'a +').

```
>> fid=fopen('magic.bin','w+')
```

fid =

3

The system assigns the ID 3 to our file, and then writes the matrix *M* to it.

```
>> fwrite(3,M)
```

ans =

16

We have written the matrix *M* to the binary file *magic.bin* of ID 3. MATLAB returns the number of elements in the file, which in this case is 16. We then close the file and the information is recorded on disk.

```
>> fclose (3)
```

ans =

0

As the answer is zero, the file was successfully closed, and the newly created file will appear in the Active Directory.

```
>> dir
```

```
.                ..              five.bin cosint.m exponen.txt id4.bin magic.bin
```

You can see the newly created file in Active Directory with its properties.

```
>> ! dir
```

```
Volume in drive C has no label.
The volume serial number £ n is: 1059-8290

Directory of C:\MATLAB6p1\work

03/01/2001 19:50 < DIR >.
03/01/2001 19:50 < DIR >...
10/06/2000 23:41 457 cosint.m
10/01/2001 22:14 64 id4.bin
10/01/2001 23:17 231 exponen.txt
11/01/2001 00: 12 10 five.bin
12/01/2001 23:09 16 magic.bin
5 files 778 bytes
2 dirs 18.027.282.432 bytes free
```

EXERCISE 3-2

Consider the identity matrix of order 4 and write it to a binary file with 32-bit floating point format. Subsequently retrieve this file and read its contents in the same array form as it was recorded. Then add to the above matrix a column of ones and save it as a binary file with the same name. Read the binary file to check its contents.

We start by generating the identity matrix of order 4:

```
>> I = eye (4)
```

I =

1 0 0 0
0 1 0 0
0 0 1 0
0 0 0 1

We open a binary file named *id4.bin*, in which we are going to save the matrix *I*, with write permission:

```
>> fid=fopen('id4.bin','w+')
```

FID =

3

We recorded the matrix *I* in the previously opened file with 32-bit floating point format:

```
>> fwrite(3,I,'float32')
```

ans =

16

Once the 16 elements of the array have been recorded, we close the file:

```
>> fclose (3)
```

ans =
0

We open it with read permission to read the contents of the previously recorded file:

```
>> fid=fopen('id4.bin','r+')
```

fid =

3

Now we read the 16 elements of the opened file in the same matrix structure and format in which it was saved.

```
>> [R,count]=fread(3,[4,4],'float32')
```

```
R =
1 0 0 0
0 1 0 0
0 0 1 0
0 0 0 1
```

```
count =
```

```
16
```

After checking the contents, we close the file:

```
>> fclose (3)
```

```
ans =
```

```
0
```

We then open the file with the proper write permission to add information without losing the existing data:

```
>> fid=fopen('id4.bin','a+')
```

```
fid =
```

```
3
```

We now add a column of ones to the end of the file's contents and close it:

```
>> fwrite(3,[1 1 1 1]','float32')
```

```
ans =
```

```
4
```

```
>> fclose(3)
```

```
ans =
```

```
0
```

Now we open the file with read permission to view its contents:

```
>> fid=fopen('id4.bin','r+')
```

```
fid =
```

```
3
```

Finally, we read the 20 items in the file in the appropriate array form and check that the column has been added to the end:

```
>> [R,count]=fread(3,[4,5],'float32')

R =

1    0    0    0    1
0    1    0    0    1
0    0    1    0    1
0    0    0    1    1

count =

20
```

EXERCISE 3-3

Generate an ASCII file named log.txt containing the values of the natural logarithm for values of the variable between 1 and 2 separated by 0.1. The format of the text in the file should consist of two columns of real floating point numbers, in such a way that the values of the variable appear in the first column and the corresponding values of the logarithm appear in the second column. Finally, display the contents of the file on screen.

```
>> x = 1:. 1:2;
y = [x; log (x)];
FID = fopen ('log. txt', 'w');
(% 12 fprintf(fid,'%6.2f. 8f\n', and);
fclose (fid)

ans =

0
```

Let us see how we can display the information directly on screen without having to save it to disk:

```
>> x = 1:. 1:2;
y = [x; log (x)];
(% 12 fprintf('%6.2f. 8f\n', and)
1.00 0.00000000
1.10 0.09531018
1.20 0.18232156
1.30 0.26236426
1.40 0.33647224
1.50 0.40546511
1.60 0.47000363
1.70 0.53062825
1.80 0.58778666
1.90 0.64185389
2.00 0.69314718
```

EXERCISE 3-4

Read the ASCII file named log.txt generated in the previous exercise. The format of the text must consist of two columns of real numbers with maximum precision in the minimum of space, so that the first column lists the values of the variable and the second column shows the corresponding values of the logarithm.

```
>> fid=fopen('log.txt');
a = fscanf(fid,'%g  %g', [2 inf]);
a = a'

a =

1.0000        0
1.1000   0.0953
1.2000   0.1823
1.3000   0.2624
1.4000   0.3365
1.5000   0.4055
1.6000   0.4700
1.7000   0.5306
1.8000   0.5878
1.9000   0.6419
2.0000   0.6931

>> fclose(fid);
```

CHAPTER 4

MATLAB Language: M-Files, Scripts, Flow Control and Numerical Analysis Functions

MATLAB and Programming

MATLAB can be used as a high-level programming language including data structures, functions, instructions for flow control, management of inputs/outputs and even object-oriented programming. MATLAB programs are usually written into files called M-files. An M-file is nothing more than a MATLAB code (*script*) that executes a series of commands or functions that accept arguments and produce an output. The M-files are created using the text editor, as described in Chapter 2.

The Text Editor

The *Editor/Debugger* is activated by clicking on the *create a new M-file* button ☐ in the MATLAB desktop or by selecting *File* ➤ *New* ➤ *M-file* in the MATLAB desktop (Figure 4-1) or Command Window (Figure 4-2). The *Editor/Debugger* opens a file in which we create the M-file, i.e. a blank file into which we will write MATLAB programming code (Figure 4-3). You can open an existing M-file using *File* ➤ *Open* on the MATLAB desktop (Figure 4-1) or, alternatively, you can use the command *Open* in the Command Window (Figure 4-2). You can also open the *Editor/Debugger* by right-clicking on the *Current Directory* window and choosing *New* ➤ *M-file* from the resulting pop-up menu (Figure 4-4). Using the menu option *Open*, you can open an existing M-file. You can open several M-files simultaneously, each of which will appear in a different window.

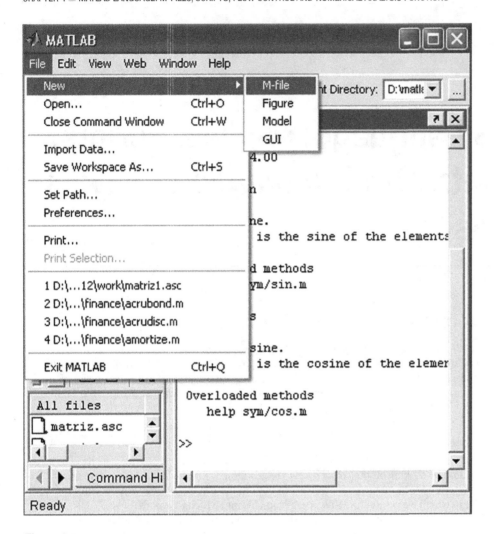

Figure 4-1.

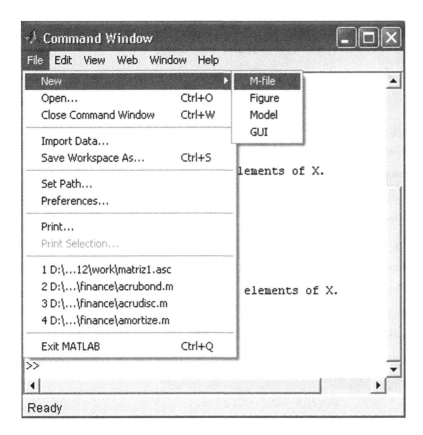

Figure 4-2.

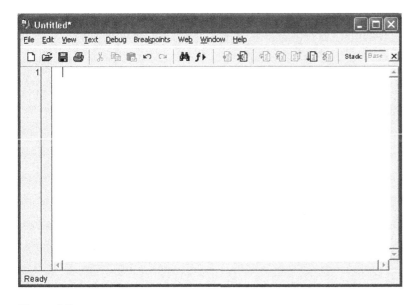

Figure 4-3.

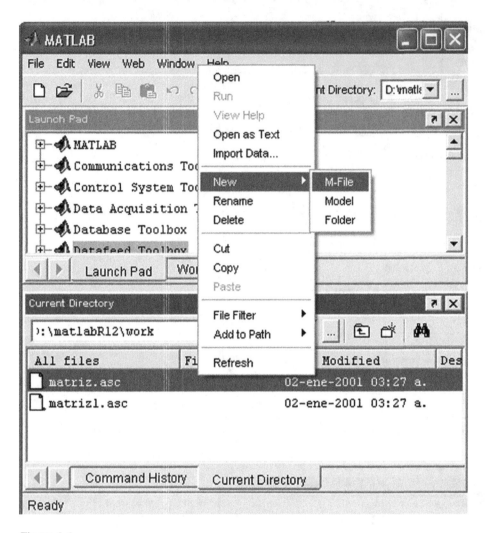

Figure 4-4.

Figure 4-5 shows the functions of the icons in the *Editor/Debugger*.

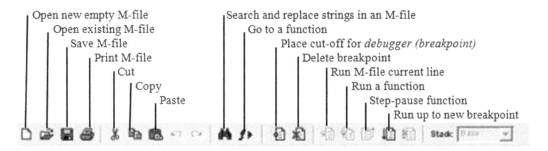

Figure 4-5.

Scripts

Scripts are the simplest possible M-files. A script has no input or output arguments. It simply consists of instructions that MATLAB executes sequentially and that could also be submitted in a sequence in the Command Window. Scripts operate with existing data on the workspace or new data created by the script. Any variable that is used by a script will remain in the workspace and can be used in further calculations after the end of the script.

Below is an example of a script that generates several curves in polar form, representing flower petals. Once the syntax of the script has been entered into the editor (Figure 4-6), it is stored in the work library (*work*) and simultaneously executes by clicking the button □ or by selecting the option *Save and run* from the *Debug* menu (or pressing F5). To move from one chart to the next press ENTER.

```
%M-file script producing graphics of petals
theta = -pi:0.01:pi;
rho(1,:) = 2*sin(5*theta).^2;
rho(2,:) = cos(10*theta).^3;
rho(3,:) = sin(theta).^2;
rho(4,:) = 5*cos(3.5*theta).^3;
for i = 1:4
    polar(theta,rho(i,:))
    pause
end
```

Figure 4-6.

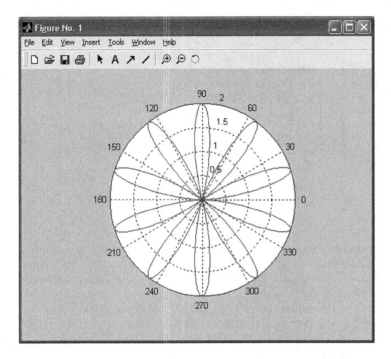

Figure 4-7.

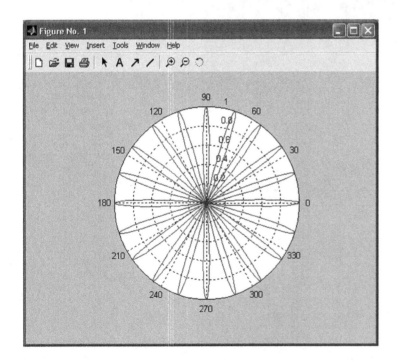

Figure 4-8.

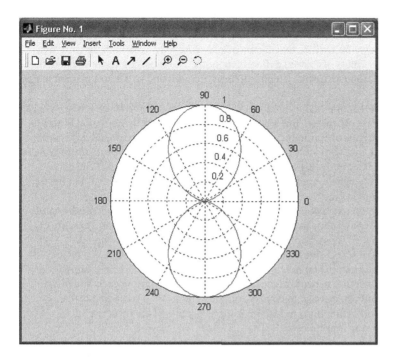

Figure 4-9.

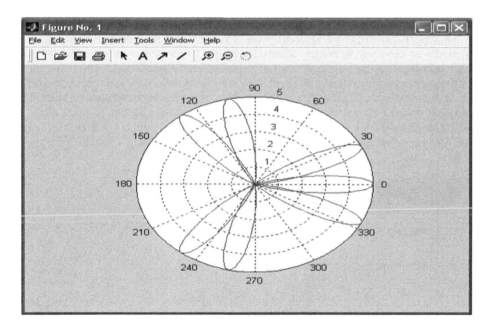

Figure 4-10.

Functions and M-files. Eval and Feval

We already know that MATLAB has a wide variety of functions that can be used in everyday work with the program. But, in addition, the program also offers the possibility of custom defined functions. The most common way to define a function is to write its definition to a text file, called an M-file, which will be permanent and will therefore enable the function to be used whenever required.

MATLAB is usually used in *command mode* (or *interactive mode*), in which case a command is written in a single line in the Command Window and is immediately processed. But MATLAB also allows the implementation of sets of commands in *batch* mode, in which case a sequence of commands can be submitted which were previously written in a file. This file (M-file) must be stored on disk with the extension ".m" in the MATLAB subdirectory, using any ASCII editor or by selecting *M-file New* from the *File* menu in the top menu bar, which opens a text editor that will allow you to write command lines and save the file with a given name. Selecting *M-File Open* from the *File* menu in the top menu bar allows you to edit any pre-existing M-file.

To run an M-file simply type its name (without extension) in interactive mode into the Command Window and press *Enter*. MATLAB sequentially interprets all commands and statements of the M-file line by line and executes them. Normally the literal commands that MATLAB is performing do not appear on screen, except when the command *echo on* is active and only the results of successive executions of the interpreted commands are displayed. Normally, work in batch mode is useful when automating large scale tedious processes which, if done manually, would be prone to mistakes. You can enter explanatory text and comments into M-files by starting each line of the comment with the symbol %. The *help* command can be used to display comments made in a particular M-file.

The command *function* allows the definition of functions in MATLAB, making it one of the most useful applications of M-files. The syntax of this command is as follows:

> function output_parameters = function_name (input_parameters)

> the function body

Once the function has been defined, it is stored in an M-file for later use. It is also useful to enter some explanatory text in the syntax of the function (using %), which can be accessed later by using the *help* command.

When there is more than one output parameter, they are placed between square brackets and separated by commas. If there is more than one input parameter, they are separated by commas. The body of the function is the syntax that defines it, and should include commands or instructions that assign values to output parameters. Each command or instruction of the body often appears in a line that ends either with a comma or, when variables are being defined, by a semicolon (in order to avoid duplication of outputs when executing the function). The function is stored in the M-file named *function_name.m*.

Let us define the function *fun1(x)* = $x ^ 3 - 2x + \cos(x)$, creating the corresponding M-file *fun1.m*. To define this function in MATLAB select *M-file New* from the *File* menu in the top menu bar (or click the button 🗋 in the MATLAB tool bar). This opens the *MATLAB Editor/Debugger* text editor that will allow us to insert command lines defining the function, as shown in Figure 4-11.

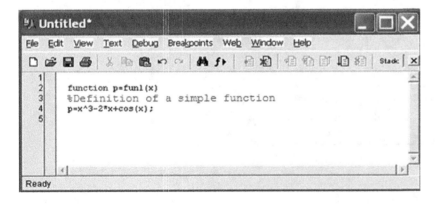

Figure 4-11.

To permanently save this code in MATLAB select the *Save* option from the *File* menu at the top of the *MATLAB Editor/Debugger*. This opens the *Save* dialog of Figure 4-12, which we use to save our function with the desired name and in the subdirectory indicated as a path in the *file name* field. Alternatively you can click on the button ⧉ or select *Save and run* from the *Debug* menu. Functions should be saved using a file name equal to the name of the function and in MATLAB's default work subdirectory *C: \MATLAB6p1\work*.

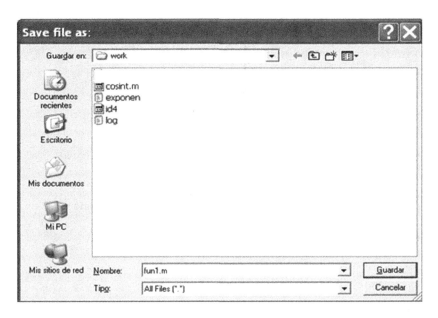

Figure 4-12.

Once a function has been defined and saved in an M-file, it can be used from the Command Window. For example, to find the value of the function at 3π-2 we write in the Command Window:

```
>> fun1(3*pi/2)
```

```
ans =
```

```
95.2214
```

For help on the previous function (assuming that comments were added to the M-file that defines it) you use the command *help*, as follows:

```
>> help fun1(x)
```

```
A simple function definition
```

A function can also be evaluated at some given arguments (input parameters) via the *feval* command, the syntax of which is as follows:

```
feval ('F', arg1, arg1,..., argn)
```

This evaluates the function F (the M-file F.m) at the specified arguments arg1, arg2,..., argn.

As an example we build an M-file named *equation2.m* which contains the function equation2, whose arguments are the three coefficients of the quadratic equation $ax^2 + bx + c = 0$ and whose outputs are the two solutions (Figure 4-13).

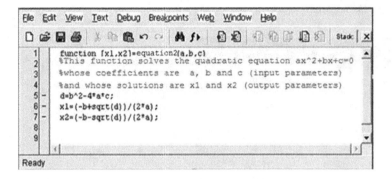

Figure 4-13.

Now if we want to solve the equation $x^2 + 2x + 3 = 0$ using *feval*, we write the following in the Command Window:

```
>> [x 1, x 2] = feval('equation2',1,2,3)

x 1 =

-1.0000 + 1. 4142i

x 2 =

-1.0000 - 1. 4142i
```

The quadratic equation can also be solved as follows:

```
>> [x 1, x 2] = equation2 (1,2,3)

x 1 =

  -1.0000 + 1. 4142i

x 2 =

-1.0000 - 1. 4142i
```

If we want to ask for help about the function equation2 we do the following:

```
>> help equation2

This function solves the quadratic equation ax ^ 2 + bx + c = 0
whose coefficients are a, b and c (input parameters)
and whose solutions are x 1 and x 2 (output parameters)
```

Evaluating a function when its arguments (input parameters) are strings is performed via the command *eval*, whose syntax is as follows:

```
eval (expression)
```

This executes the expression when it is a string.

As an example, we evaluate a string that defines a magic square of order 4.

```
>> n=4;
>> eval(['M' num2str(n) ' = magic(n)'])

M4 =

16  2   3   13
 5  11  10  8
 9  7   6   12
 4  14  15  1
```

Local and Global Variables

Typically, each function defined as an M-file contains local variables, i.e., variables that have effect only within the M-file, separate from other M-files and the base workspace. However, it is possible to define variables inside M-files which can take effect simultaneously in other M-files and in the base workspace. For this purpose, it is necessary to define global variables with the GLOBAL command whose syntax is as follows:

```
GLOBAL x y z...
```

This defines the variables x, y and z as global.

Any variables defined as global inside a function are available separately for the rest of the functions and in the base workspace command line. If a global variable does not exist, the first time it is used, it will be initialized as an empty array. If there is already a variable with the same name as a global variable being defined, MATLAB will send a warning message and change the value of that variable to match the global variable. It is convenient to declare a variable as global in every function that will need access to it, and also in the command line, in order to access it from the base workspace. The GLOBAL command is located at the beginning of a function (before any occurrence of the variable).

As an example, suppose that we want to study the effect of the interaction coefficients α and β in the Lotka–Volterra predator-prey model:

$$\dot{y}_1 = y_1 - \alpha y_1 y_2$$
$$\dot{y}_2 = -y_2 - \beta y_1 y_2$$

To do this, we create the function *lotka* in the M-file *lotka.m* as depicted in Figure 4-14.

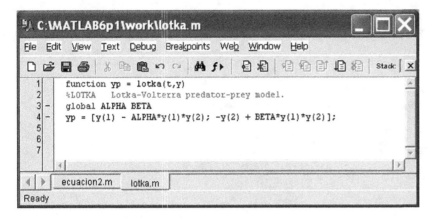

Figure 4-14.

Later, we might type the following in the command line:

```
>> global ALPHA BETA
ALPHA = 0.01
BETA  = 0.02
```

These global values may then be used for α and β in the M-file *lotka.m* (without having to specify them). For example, we can generate the graph (Figure 4-15) with the following syntax:

```
>> [t, y] = ode23 ('lotka', 0.10, [1; 1]); plot(t,y)
```

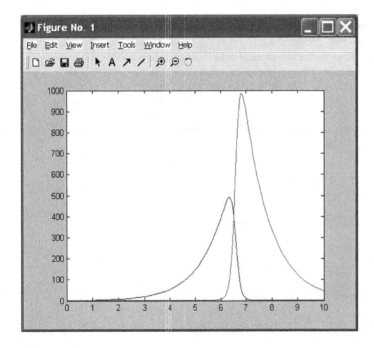

Figure 4-15.

Data Types

MATLAB has 14 different data types, summarized in Figure 4-16 below.

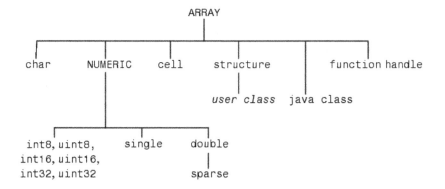

Figure 4-16.

Below are the different types of data:

Data type	Example	Description
single	3* 10 ^ 38	*Simple numerical precision. This requires less storage than double precision, but it is less precise. This type of data should not be used in mathematical operations.*
Double	3*10^300 5+6i	*Double numerical precision. This is the most commonly used data type in MATLAB.*
sparse	speye(5)	*Sparse matrix with double precision.*
int8, uint8, int16, uint16, int32, uint32	UInt8(magic (3))	*Integers and unsigned integers with 8, 16, and 32 bits. These make it possible to use entire amounts with efficient memory management. This type of data should not be used in mathematical operations.*
char	'Hello'	*Characters (each character has a length of 16 bits).*
cell	{17 'hello' eye (2)}	*Cell (contains data of similar size).*
structure	a.day = 12; a.color = 'Red'; a.mat = magic(3);	*Structure (contains cells of similar size).*
user class	inline('sin (x)')	*MATLAB class (built with functions).*
java class	Java. awt.Frame	*Java class (defined in API or own) with Java.*
function handle	@humps	*Manages functions in MATLAB. It can be last in a list of arguments and evaluated with feval.*

Flow Control: FOR Loops, WHILE and IF ELSEIF

The use of recursive functions, conditional operations and piecewise defined functions is very common in mathematics. The handling of loops is necessary for the definition of these types of functions. Naturally, the definition of the functions will be made via *M-files*.

FOR Loops

MATLAB has its own version of the DO statement (defined in the syntax of most programming languages). This statement allows you to run a command or group of commands repeatedly. For example:

» for i=1:3, x(i)=0, end

X =

0

X =

0   0

X =

0 0 0

The general form of a FOR loop is as follows:

```
for variable = expression
    commands
end
```

The loop always starts with the clause *for* and ends with the clause *end*, and includes in its interior a whole set of commands that are separated by commas. If any command defines a variable, it must end with a semicolon in order to avoid repetition in the output. Typically, loops are used in the syntax of M-files. Here is an example (Figure 4-17):

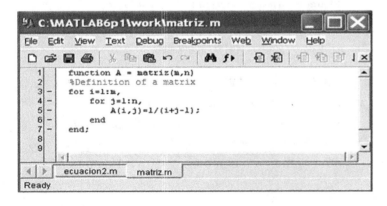

Figure 4-17.

In this loop we have defined a Hilbert matrix of order *(m, n)*. If we save it as an M-file *matriz.m*, we can build any Hilbert matrix later by running the M-file and specifying values for the variables *m* and *n* (the matrix dimensions) as shown below:

```
>> M = matriz (4,5)
```

M =

```
1.0000 0.5000 0.3333 0.2500 0.2000
0.5000 0.3333 0.2500 0.2000 0.1667
0.3333 0.2500 0.2000 0.1667 0.1429
0.2500 0.2000 0.1667 0.1429 0.1250
```

WHILE Loops

MATLAB has its own version of the WHILE structure defined in the syntax of most programming languages. This statement allows you to repeat a command or group of commands a number of times while a specified logical condition is met. The general syntax of this loop is as follows:

```
While condition
    commands
end
```

The loop always starts with the clause *while*, followed by a condition, and ends with the clause *end*, and includes in its interior a whole set of commands that are separated by commas which continually loop while the condition is met. If any command defines a variable, it must end with a semicolon in order to avoid repetition in the output. As an example, we write an M-file (Figure 4-18) that is saved as *while1.m*, which calculates the largest number whose factorial does not exceed 10^{100}.

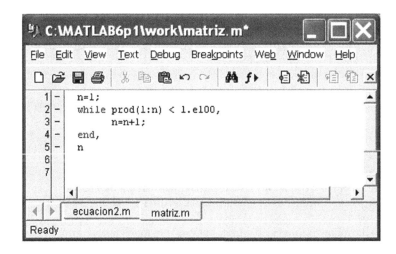

Figure 4-18.

We now run the M-file.

```
>> while1

n =

70
```

IF ELSEIF ELSE END Loops

MATLAB, like most structured programming languages, also includes the IF-ELSEIF-ELSE-END structure. Using this structure, scripts can be run if certain conditions are met. The loop syntax is as follows:

```
if condition
    commands
end
```

In this case the commands are executed if the condition is true. But the syntax of this loop may be more general.

```
if condition
    commands1
else
    commands2
end
```

In this case, the commands *commands1* are executed if the condition is true, and the commands *commands2* are executed if the condition is false.

IF statements and FOR statements can be nested. When multiple IF statements are nested using the ELSEIF statement, the general syntax is as follows:

```
if condition1
    commands1
elseif condition2
    commands2
elseif condition3
    commands3
    .
    .
    .
else
end
```

In this case, the commands *commands1* are executed if c*ondition1* is true, the commands *commands2* are executed if *condition1* is false and *condition2* is true, the commands *commands3* are executed if *condition1* and *condition2* are false and *condition3* is true, and so on.

The previous nested syntax is equivalent to the following unnested syntax, but executes much faster:

```
if condition1
    commands1
else
        if condition2
            commands2
```

136

```
     else
         if condition3
             commands3
         else
         .
         .
         .
         end
     end
end
```

Consider, for example, the M-file *else1.m* (see Figure 4-19).

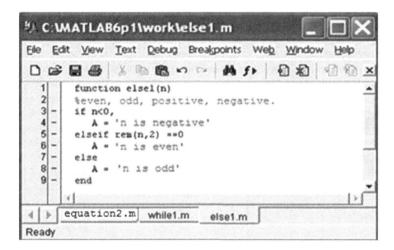

Figure 4-19.

When you run the file it returns negative, odd or even according to whether the argument *n* is negative, non-negative and odd, or non-negative and even, respectively:

>> **else1 (8), else1 (5), else1 (- 10)**

A =

n is even

A =

n is odd

A =

n is negative

Switch and Case

The *switch* statement executes certain statements based on the value of a variable or expression. Its basic syntax is as follows:

```
switch expression (scalar or string)
case value1
statements % runs if expression is value1
case value2
statements % runs if expression is value2
.
.
.
otherwise
statements % runs if neither case is satisfied

end
```

Below is an example of a function that returns 'minus one', 'zero', 'one', or 'another value' according to whether the input is equal to –1,0,1 or something else, respectively (Figure 4-20).

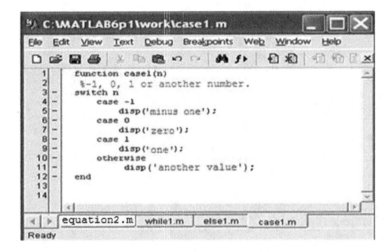

Figure 4-20.

Running the above example we get:

```
>> case1 (25)
another value

>> case1 (- 1)
minus one
```

Continue

The *continue* statement passes control to the next iteration in a *for* loop or *while* loop in which it appears, ignoring the remaining instructions in the body of the loop. Below is an M-file *continue.m* (Figure 4-21) that counts the lines of code in the file *magic.m*, ignoring the white lines and comments.

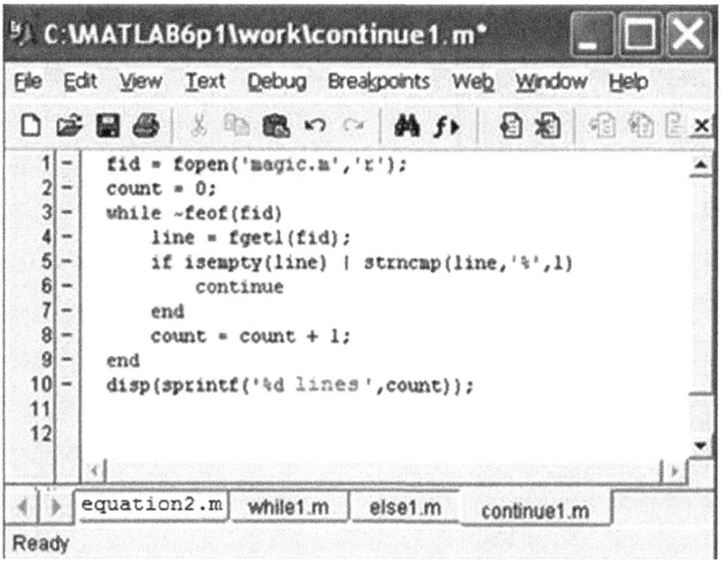

Figure 4-21.

Running the M-file, we get:

```
>> continue1
25 lines
```

Break

The *break* statement terminates the execution of a *for* loop or *while* loop, skipping to the first instruction which appears outside of the loop. Below is an M-file *break1.m* (Figure 4-22) which reads the lines of code in the file *fft.m*, exiting the loop as soon as it encounters the first empty line.

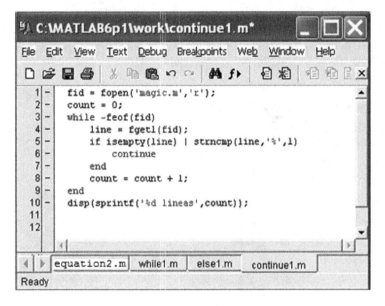

```
fid = fopen('magic.m','r');
count = 0;
while ~feof(fid)
    line = fgetl(fid);
    if isempty(line) | strncmp(line,'%',1)
        continue
    end
    count = count + 1;
end
disp(sprintf('%d lineas',count));
```

Figure 4-22.

Running the M-file we get:

>> break1

```
%FFT Discrete Fourier transform.
%   FFT(X) is the discrete Fourier transform (DFT) of vector X.  For
%   matrices, the FFT operation is applied to each column. For N-D
%   arrays, the FFT operation operates on the first non-singleton
%   dimension.
%
%   FFT(X,N) is the N-point FFT, padded with zeros if X has less
%   than N points and truncated if it has more.
%
%   FFT(X,[],DIM) or FFT(X,N,DIM) applies the FFT operation across the
%   dimension DIM.
%
%   For length N input vector x, the DFT is a length N vector X,
%   with elements
%                     N
%     X(k) =         sum  x(n)*exp(-j*2*pi*(k-1)*(n-1)/N), 1 <= k <= N.
%                    n=1
%   The inverse DFT (computed by IFFT) is given by
%                     N
%     x(n) = (1/N) sum  X(k)*exp( j*2*pi*(k-1)*(n-1)/N), 1 <= n <= N.
%                    k=1
%
%   See also IFFT, FFT2, IFFT2, FFTSHIFT.
```

Try... Catch

The instructions between *try* and *catch* are executed until an error occurs. The instruction *lasterr* is used to show the cause of the error. The general syntax of the command is as follows:

```
try,
instruction
...,
instruction
catch,
instruction
...,
instruction
end
```

Return

The *return* statement terminates the current script and returns the control to the invoked function or the keyboard. The following is an example (Figure 4-23) that computes the determinant of a non-empty matrix. If the array is empty it returns the value 1.

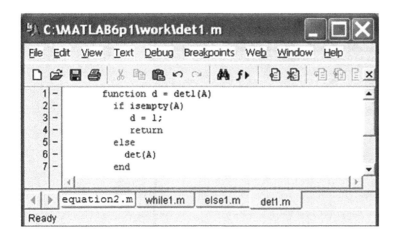

Figure 4-23.

Running the function for a non-empty array we get:

```
>> A = [- 1, - 1, 1; 1,0,1; 1,1,1]

A =
-1 -1 -1
 1  0  1
 1 -1 -1

>> det1 (A)

ans =

2
```

Now we apply the function to an empty array:

```
>> B =[]

B =

    []

>> det1 (B)

ans =

    1
```

Subfunctions

M-file-defined functions can contain code for more than one function. The main function in an M-file is called a *primary function*, which is precisely the function which invokes the M-file, but subfunctions hanging from the primary function may be added which are only visible for the primary function or another subfunction within the same M-file. Each subfunction begins with its own function definition. An example is shown in Figure 4-24.

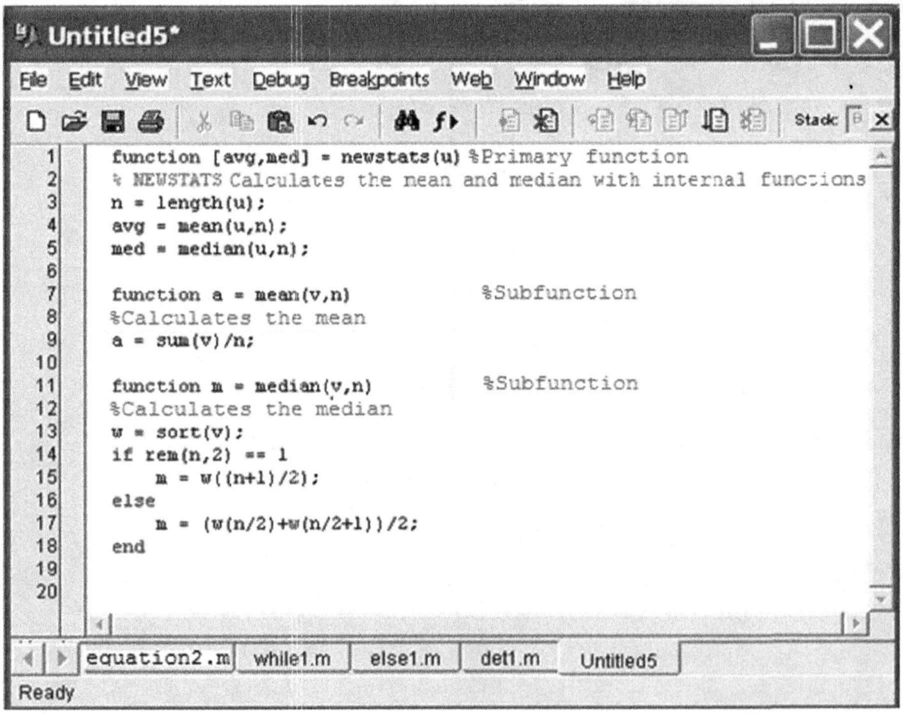

Figure 4-24.

The subfunctions *mean* and *median* calculate the arithmetic mean and the median of the input list. The primary function *newstats* determines the length *n* of the list and calls the subfunctions with the list as the first argument and *n* as the second argument. When executing the main function, it is enough to provide as input a list of values for which the arithmetic mean and median will be calculated. The subfunctions are executed automatically, as shown below.

```
>> [mean, median] = newstats ([10,20,3,4,5,6])

mean =

    8

median =

    5.5000
```

Commands in M-files

MATLAB provides certain procedural commands which are often used in M-file scripts. Among them are the following:

echo on	*View on-screen commands of an M-file script while it is running.*
echo off	*Hides on-screen commands of an M-file script (this is the default setting).*
pause	*Interrupts the execution of an M-file until the user presses a key to continue.*
pause(n)	*Interrupts the execution of an M-file for n seconds.*
pause off	*Disables pause and pause (n).*
pause on	*Enables pause and pause (n).*
keyboard	*Interrupts the execution of an M-file and passes the control to the keyboard so that the user can perform other tasks. The execution of the M-file can be resumed by typing the* return *command into the Command Window and pressing Enter.*
return	*Resumes execution of an M-file after an outage.*
break	*Prematurely exits a loop.*
CLC	*Clears the Command Window.*
Home	*Hides the cursor.*
more on	*Enables paging of the MATLAB Command Window output.*
more off	*Disables paging of the MATLAB Command Window output.*
more (N)	*Sets page size to N lines.*
menu	*Offers a choice between various types of menu for user input.*

Functions Relating to Arrays of Cells

An array is a well-ordered collection of individual items. This is simply a list of elements, each of which is associated with a positive integer called its index, which represents the position of that element in the list. It is essential that each element is associated with a unique index, which can be zero or negative, which identifies it fully, so that to make changes to any elements of the array it suffices to refer to their indices. Arrays can be of one or more dimensions, and correspondingly they have one or more sets of indices that identify their elements. The most important commands and functions that enable MATLAB to work with arrays of cells are the following:

c = cell(n)	*Creates an n×n array whose cells are empty arrays.*
c = cell(m,n)	*Creates an m×n array whose cells are empty arrays.*
c = cell([m n])	*Creates an m×n array whose cells are empty arrays.*
c = cell(m,n,p,...)	*Creates an m×n×p×... array of empty arrays.*
c = cell([m n p ...])	*Creates an m×n×p×... array of empty arrays.*
c = cell(size(A))	*Creates an array of empty arrays of the same size as A.*
D = cellfun('f',C)	*Applies the function f (isempty, islogical, isreal, length, ndims, or prodofsize) to each element of the array C.*
D = cellfun('size',C,k)	*Returns the size of each element of dimension k in C.*
D = cellfun('isclass',C,class)	*Returns true for each element of C corresponding to class.*
C=cellstr(S)	*Places each row of the character array S into separate cells of C.*
S = cell2struct(C,fields,dim)	*Converts the array C to a structure array S incorporating field names 'fields' and the dimension 'dim' of C.*
celldisp (C)	*Displays the contents of the array C.*
celldisp(C, name)	*Assigns the contents of the array C to the variable name.*
cellplot(C)	*Shows a graphical representation of the array C.*
cellplot(C,'legend')	*Shows a graphical representation of the array C and incorporates a legend.*
C = num2cell(A)	*Converts a numeric array A to the cell array C.*
C = num2cell(A,dims)	*Converts a numeric array A to a cell array C placing the given dimensions in separate cells.*

As a first example, we create an array of cells of the same size as the unit square matrix of order two.

```
>> A = ones(2,2)

A =
1    1
1    1

>> c = cell(size(A))

C =

[]    []
[]    []
```

We then define and present a 2 × 3 array of cells element by element, and apply various functions to the cells.

```
>> C {1.1} = [1 2; 4 5];
C {1,2} = 'Name';
C {1,3} = pi;
C{2,1} = 2 + 4i;
C{2,2} = 7;
C{2,3} = magic(3);

>> C

C =

[2x2 double]        'Name'     [    3.1416]
[2.0000+ 4.0000i]    [   7]     [3x3 double]
```

```
>> D = cellfun('isreal',C)

D =

1    1    1
0    1    1
```

```
>> len = cellfun('length',C)

len =

2    4    1
1    1    3
```

```
>> isdbl = cellfun('isclass',C,'double')

isdbl =

1 0 1
1 1 1
```

The contents of the cells in the array C defined above are revealed using the command *celldisp*.

```
>> celldisp(C)

C{1,1} =
1    2
4    5

C{2,1} =

2.0000 + 4.0000i
```

C{1,2} =

Name

C {2,2} =

7

C {1,3} =

3.1416

C {2,3} =

8 1 6
3 5 7
4 9 2

The following displays a graphical representation of the array C (Figure 4-25).

>> cellplot(C)

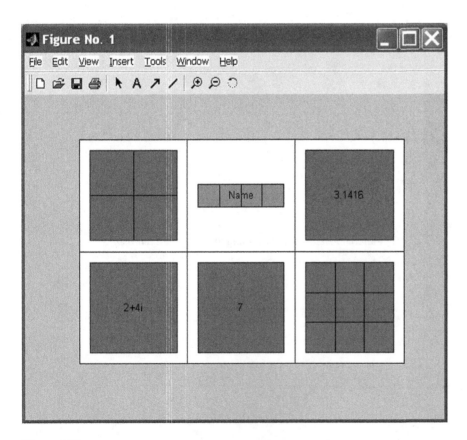

Figure 4-25.

Multidimensional Array Functions

The following group of functions is used by MATLAB to work with multidimensional arrays:

C = **cat(dim,A,B)**	*Concatenates arrays A and B according to the dimension dim.*
C = **cat(dim,A1,A2,A3,A4...)**	*Concatenates arrays A1, A2,... according to the dimension dim.*
B = **flipdim (A, dim)**	*Flips the array A along the specified dimension dim.*
[I,J] = **ind2sub(*siz*,IND)**	*Returns the matrices I and J containing the equivalent row and column subscripts corresponding to each index in the matrix IND for a matrix of size siz.*
[I1,I2,I3,...,In] = **ind2sub(*siz*,IND)**	*Returns matrices I1, I2,...,In containing the equivalent row and column subscripts corresponding to each index in the matrix IND for a matrix of size siz.*
A = **ipermute(B,*order*)**	*Inverts the dimensions of the multidimensional array D according to the values of the vector order.*
[X1, X2, X3,...] = **ndgrid(x1,x2,x3,...)**	*Transforms the domain specified by vectors x1, x2,... into the arrays X1, X2,... which can be used for evaluation of functions of several variables and interpolation.*
[X 1, X 2,...] = **ndgrid (x)**	*Equivalent to ndgrid(x,x,x,...).*
n = **ndims(A)**	*Returns the number of dimensions in the array A.*
B = **permute(A,order)**	*Swaps the dimensions of the array A specified by the vector order.*
B = **reshape(A,m,n)**	*Defines an m×n matrix B whose elements are the columns of a.*
B = **reshape(A,m,n,p,...)**	*Defines an array B whose elements are those of the array A restructured according to the dimensions m×n×p×...*
B = **reshape(A,[m n p...])**	*Equivalent to B = reshape(A,m,n,p,...)*
B = **reshape(A,siz)**	*Defines an array B whose elements are those of the array A restructured according to the dimensions of the vector siz.*
B = **shiftdim(X,n)**	*Shifts the dimensions of the array X by n, creating a new array B.*
[B,nshifts] = **shiftdim(X)**	*Defines an array B with the same number of elements as X but with leading singleton dimensions removed.*
B=**squeeze(A)**	*Creates an array B with the same number of elements as A but with all singleton dimensions removed.*
IND = **sub2ind(siz,I,J)**	*Gives the linear index equivalent to the row and column indices I and J for a matrix of size siz.*
IND = **sub2ind(siz,I1,I2,...,In)**	*Gives the linear index equivalent to the n indices I1, I2,..., in a matrix of size siz.*

As a first example we concatenate a magic square and Pascal matrix of order 3.

```
>> A = magic (3); B = pascal (3);
>> C = cat (4, A, B)

C(:,:,1,1) =

8 1 6
3 5 7
4 9 2

C(:,:,1,2) =

1 1 1
1 2 3
1 3 6
```

The following example flips the Rosser matrix.

```
>> R=rosser

R =

    611    196   -192    407     -8    -52    -49     29
    196    899    113   -192    -71    -43     -8    -44
   -192    113    899    196     61     49      8     52
    407   -192    196    611      8     44     59    -23
     -8    -71     61      8    411   -599    208    208
    -52    -43     49     44   -599    411    208    208
    -49     -8      8     59    208    208     99   -911
     29    -44     52    -23    208    208   -911     99
```

```
>> flipdim(R,1)

ans =

ans =

     29    -44     52    -23    208    208   -911     99
    -49     -8      8     59    208    208     99   -911
    -52    -43     49     44   -599    411    208    208
     -8    -71     61      8    411   -599    208    208
    407   -192    196    611      8     44     59    -23
   -192    113    899    196     61     49      8     52
    196    899    113   -192    -71    -43     -8    -44
    611    196   -192    407     -8    -52    -49     29
```

Now we define an array by concatenation and permute and inverse permute its elements.

```
>> a = cat(3,eye(2),2*eye(2),3*eye(2))

a(:,:,1) =

1 0
0 1

a(:,:,2) =

2 0
0 2

a(:,:,3) =

3 0
0 3

>> B = permute(a,[3 2 1])

B(:,:,1) =

1 0
2 0
3 0

B(:,:,2) =

0 1
0 2
0 3

>> C = ipermute(B,[3 2 1])

C(:,:,1) =

1 0
0 1

C(:,:,2) =

2 0
0 2

C(:,:,3) =

3 0
0 3
```

The following example evaluates the function $f(x_1, x_2) = x_1 e^{-x_1^2 - x_2^2}$ in the square $[-2, 2] \times [-2, 2]$ and displays it graphically (Figure 4-26).

```
>> [X 1, X 2] = ndgrid(-2:.2:2,-2:.2:2);
Z = X 1. * exp(-X1.^2-X2.^2);
mesh (Z)
```

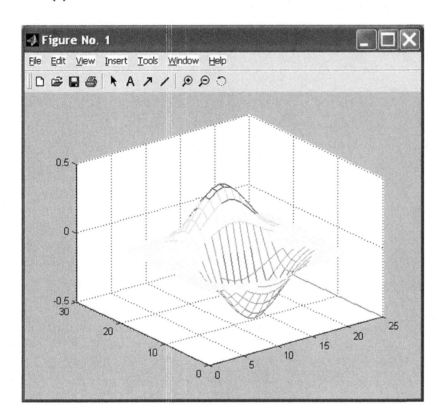

Figure 4-26.

In the following example we resize a 3×4 random matrix to a 2×6 matrix.

```
>> A=rand(3,4)
```

A =

```
0.9501    0.4860    0.4565    0.4447
0.2311    0.8913    0.0185    0.6154
0.6068    0.7621    0.8214    0.7919
```

```
>> B = reshape(A,2,6)
```

B =

```
0.9501 0.6068 0.8913 0.4565 0.8214 0.6154
0.2311 0.4860 0.7621 0.0185 0.4447 0.7919
```

Numerical Analysis Methods in MATLAB

MATLAB programming techniques allow you to implement a wide range of numerical algorithms. It is possible to design programs which perform numerical integration and differentiation, solve differential equations, optimize non-linear functions, etc. However, MATLAB's Basic module already has a number of tailor-made functions which implement some of these algorithms. These functions are set out in the following subsections. In the next chapter we will give some examples showing how these functions can be used in practice.

Zeros of Functions and Optimization

The commands (functions) that enables MATLAB's Basic module to optimize functions and find the zeros of functions are as follows:

x = fminbnd(fun,x1,x2)	*Minimizes the function on the interval (x1 x2).*
x = fminbnd(fun,x1,x2,options)	*Minimizes the function on the interval (x1 x2) according to the option given by optimset (...). This last command is explained later.*
x = fminbnd(fun,x1,x2, options,P1,P2,...)	*Specifies additional parameters P1, P2, ... to pass to the target function fun(x,P1,P2, ...).*
[x, fval] = fminbnd (...)	*Returns the value of the objective function at x.*
[x, fval, f] = fminbnd (...)	*In addition, returns an indicator of convergence f (f > 0 indicates convergence to the solution, f < 0 indicates no convergence and f = 0 indicates the algorithm exceeded the maximum number of iterations).*
[x,fval,f,output] = fminbnd(...)	*Provides further information (output.algorithm gives the algorithm used, output.funcCount gives the number of evaluations of fun and output.iterations gives the number of iterations).*
x = fminsearch(fun,x0)	*Returns the minimum of a scalar function of several variables, starting at an initial estimate x0. The argument x0 can be an interval [a, b].*
x = fminsearch(fun,x0,options)	*To find the minimum of fun in [a, b], x = fminsearch (fun, [a, b]) is used.*
x = fminsearch(fun,x0,options,P1,P2,...)	
[x,fval] = fminsearch(...)	
[x,fval,f] = fminsearch(...)	
[x,fval,f,output] = fminsearch(...)	
x = fzero(fun,x0)	*Finds zeros of the function fun, with initial estimate x0, by finding a point where fun changes sign. The argument x0 can be an interval [a, b]. Then, to find a zero of fun in [a, b], we use x = fzero (fun, [a, b]), where fun has opposite signs at a and b.*
x = fzero(fun,x0,options)	
x = fzero(fun,x0,options,P1,P2,...)	
[x, fval] = fzero (...)	
[x, fval, exitflag] = fzero (...)	
[x,fval,exitflag,output] = fzero(...)	

(continued)

options = optimset('p1','v1','p2','v2,...)	*Creates optimization parameters p1, p2,... with values v1, v2... The possible parameters are* `Display` *(with possible values 'off', 'iter', 'final', 'notify') to respectively not display the output, display the output of each iteration, display only the final output, and display a message if there is no convergence);* `MaxFunEvals`*, whose value is an integer indicating the maximum number of evaluations;* `MaxIter` *whose value is an integer indicating the maximum number of iterations;* `TolFun`*, whose value is an integer indicating the tolerance in the value of the function, and* `TolX`*, whose value is an integer indicating the tolerance in the value of x.*
val = optimget (options, 'param')	*Returns the value of the parameter specified in the optimization options structure.*
g = inline (*expr*)	*Transforms the string expr into a function.*
g = inline(*expr,arg1,arg2, ...*)	*Transforms the string expr into a function with given input arguments.*
g = inline (*expr, n*)	*Transforms the string expr into a function with n input arguments.*
f = @function	*Enables the function to be evaluated.*

As a first example we find the value of x that minimizes the function $\cos(x)$ in the interval $(3,4)$.

```
>> x = fminbnd(@cos,3,4)

x =
3.1416
```

We could also have used the following syntax:

```
>> x = fminbnd(inline('cos(x)'),3,4)

x =
3.1416
```

In the following example we find the above minimum to 8 decimal places and find the value of x that minimizes the cosine in the given interval, presenting information relating to all iterations of the process.

```
>> [x,fval,f] = fminbnd(@cos,3,4,optimset('TolX',1e-8,... 'Display','iter'));
```

Func-count	x	f(x)	Procedure
1	3.38197	-0.971249	initial
2	3.61803	-0.888633	golden
3	3.23607	-0.995541	golden
4	3.13571	-0.999983	parabolic
5	3.1413	-1	parabolic
6	3.14159	-1	parabolic
7	3.14159	-1	parabolic
8	3.14159	-1	parabolic
9	3.14159	-1	parabolic

```
Optimization terminated successfully:
the current x satisfies the termination criteria using OPTIONS.TolX of 1.000000e-008
```

In the following example, taking $(-1, 2; 1)$ as initial values, we find the minimum and target value of the following function of two variables:

$$f(x) = 100\left(x_2 - x^2_1\right)^2 + \left(1 - x_1\right)^2$$

```
>> [x,fval] = fminsearch(inline('100*(x(2)-x(1)^2)^2+...
(((1-x (1)) ^ 2'), [- 1.2, 1])

X =

1.0000 1.0000

fval =

8. 1777e-010
```

The following example computes a zero of the sine function with an initial estimate of 3, and a zero of the cosine function between 1 and 2.

```
>> x = fzero(@sin,3)

X =

3.1416

>> x = fzero(@cos,[1 2])

X =

    1.5708
```

Numerical Integration

MATLAB contains functions that allow you to perform numerical integration using Simpson's method and Lobato's method. The syntax of these functions is as follows:

q = **quad(f,a,b)**	*Finds the integral of f between a and b by Simpson's method with an error of 10-6.*
q = **quad(f,a,b,tol)**	*Find the integral of f between a and b by Simpson's method with the tolerance tol instead of 10-6.*
q = **quad(f,a,b,tol,trace)**	*Find the integral of f between a and b by Simpson's method with the tolerance tol and presents the trace of iterations.*
q = **quad(f,a,b,tol,trace,p1,p2, ...)**	*Passes additional arguments p1, p2, ... to the function f, f(x,p1,p2, ...).*
[q, fcnt] = **quadl(f,a,b,...)**	*Additionally returns the number of evaluations of f.*

(continued)

q = quadl(f,a,b)	*Finds the integral off between a and b by Lobato's quadrature method with a 10-6 error.*
q = quadl(f,a,b,tol)	*Finds the integral of f between a and b by Lobato's quadrature method with the tolerance tol instead of 10^{-6}.*
q = quadl(f,a,b,tol,trace)	*Finds the integral of f between a and b by Lobato's quadrature method with the tolerance tol and presents the trace of iterations.*
q = quad(f,a,b,tol,trace,p1,p2,...)	*Passes additional arguments p1, p2,... to the function f, f(x,p1,p2,...).*
[q, fcnt] = quadl(f,a,b,...)	*Additionally returns the number of evaluations off.*
q = dblquad (f, xmin, xmax, ymin, ymax)	*Evaluates the double integral f(x,y) in the rectangle specified by the given parameters, with an error of 10^{-6}. dblquad will be removed in future releases and replaced by integral2.*
q = dblquad (f, xmin, xmax, ymin,ymax,tol)	*Evaluates the double integral f(x,y) in the rectangle specified by the given parameters, with tolerance tol.*
q = dblquad (f, xmin, xmax, ymin,ymax,tol,@quadl)	*Evaluates the double integral f(x,y) in the rectangle specified by the given parameters, with tolerance tol and using the quadl method.*
q = dblquad (f, xmin, xmax, ymin,ymax,tol,method,p1,p2,...)	*Passes additional arguments p1, p2,... to the function f.*

As a first example we calculate $\int_0^2 \frac{1}{x^3 - 2x - 5} dx$ using Simpson's method.

```
>> F = inline('1./(x.^3-2*x-5)');
>> Q = quad(F,0,2)

Q =

-0.4605
```

Then we observe that the integral remains unchanged even if we increase the tolerance to 10^{-18}.

```
>> Q = quad(F,0,2,1.0e-18)

Q =

-0.4605
```

In the following example we evaluate the same integral using Lobato's method.

```
>> Q = quadl(F,0,2)

Q =

-0.4605
```

We evaluate the double integral $\int_\pi^{2\pi} \int_0^\pi (y\sin(x) + x\cos(y)) dy dx$.

```
>> Q = dblquad (inline (' y * sin (x) + x * cos (y)'), pi, 2 * pi, 0, pi)
```

Q =

   -9.8696

Numerical Differentiation

The derivative $f'(x)$ of a function $f(x)$ can simply be defined as the rate of change of $f(x)$ with respect to x. The derivative can be expressed as a ratio between the change in $f(x)$, denoted by $df(x)$, and the change in x, denoted by dx. The derivative of a function f at the point x_k can be estimated by using the expression:

$$f'(x_k) = \frac{f(x_k) - f(x_{k-1})}{x_k - x_{k-1}}$$

provided the values x_k, x_{k-1} are close to each other. Similarly the second derivative $f''(x)$ of the function $f(x)$ can be estimated as the first derivative of $f'(x)$, i.e.:

$$f''(x_k) = \frac{f'(x_k) - f'(x_{k-1})}{x_k - x_{k-1}}$$

MATLAB includes in its Basic module the function *diff*, which allows you to approximate derivatives. The syntax is as follows:

Y = diff(X) *Calculates the differences between adjacent elements in the vector X:[X(2) - X(1), X(3) - X (2),..., X(n) - X(n-1)]. If X is an m×n matrix, diff (X) returns the array of differences by rows: [X(2:m,:)-X(1:m-1,:)]*

Y = diff(X,n) *Finds differences of order n, for example: diff(X,2) = diff (diff (X)).*

As an example we consider the function $f(x) = x^5 - 3x^4 - 11x^3 + 27x^2 + 10x - 24$, find the difference vector of $[-4, -3.9, -3.8, ..., 4.8, 4.9, 5]$ the difference vector of $[f(-4), f(-3.9), f(-3.8), ..., f(4.8), f(4.9), f(5)]$ and the elementwise quotient of the latter by the former, and graph the function in the interval $[-4.5]$. See Figure 4-27.

```
>> x =-4:0.1: 5;
>> f = x.^5-3*x.^4-11*x.^3 + 27*x.^2 + 10*x-24;
>> df=diff(f)./diff(x)
```

df =

  1.0e+003 *

  Columns 1 through 7

    1.2390   1.0967   0.9655   0.8446   0.7338   0.6324   0.5400

  Columns 8 through 14

    0.4560   0.3801   0.3118   0.2505   0.1960   0.1477   0.1053

155

```
  Columns 15 through 21

    0.0683     0.0364     0.0093    -0.0136    -0.0324    -0.0476    -0.0594

  Columns 22 through 28

   -0.0682    -0.0743    -0.0779    -0.0794    -0.0789    -0.0769    -0.0734

  Columns 29 through 35

   -0.0687    -0.0631    -0.0567    -0.0497    -0.0424    -0.0349    -0.0272

  Columns 36 through 42

   -0.0197    -0.0124    -0.0054     0.0012     0.0072     0.0126     0.0173

  Columns 43 through 49

    0.0212     0.0244     0.0267     0.0281     0.0287     0.0284     0.0273

  Columns 50 through 56

    0.0253     0.0225     0.0189     0.0147     0.0098     0.0044    -0.0014

  Columns 57 through 63

   -0.0076    -0.0140    -0.0205    -0.0269    -0.0330    -0.0388    -0.0441

  Columns 64 through 70

   -0.0485    -0.0521    -0.0544    -0.0553    -0.0546    -0.0520    -0.0472

  Columns 71 through 77

-0.0400    -0.0300    -0.0170    -0.0007     0.0193     0.0432     0.0716

  Columns 78 through 84

    0.1046     0.1427     0.1863     0.2357     0.2914     0.3538     0.4233

  Columns 85 through 90

    0.5004     0.5855     0.67910.7816     0.8936     1.0156

>> plot (x, f)
```

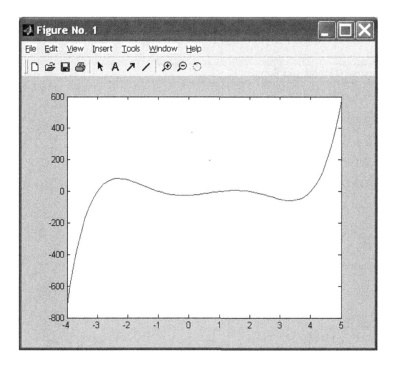

Figure 4-27.

Approximate Solution of Differential Equations

MATLAB provides commands in its Basic module allowing for the numerical solution of ordinary differential equations (ODEs), differential algebraic equations (DAEs) and boundary value problems. It is also possible to solve systems of differential equations with boundary values and parabolic and elliptic partial differential equations.

Ordinary Differential Equations with Initial Values

An ordinary differential equation contains one or more derivatives of the dependent variable y with respect to the independent variable t. A first order ordinary differential equation with an initial value for the independent variable can be represented as:

$$y' = f(t, y)$$
$$y(t_0) = y_0$$

The previous problem can be generalized to the case where y is a vector, $y = (y_1, y_2, \ldots, yn)$

MATLAB's Basic module commands relating to ordinary differential equations and differential algebraic equations with initial values are presented in the following table:

Command	Class of Problem Solving, Numerical Method and Syntax
ode45	*Ordinary differential equations by the Runge–Kutta method*
ode23	*Ordinary differential equations by the Runge–Kutta method*
ode113	*Ordinary differential equations by Adams' method*
ode15s	*Differential algebraic equations and ordinary differential equations using NDFs (BDFs)*
ode23s	*Ordinary differential equations by the Rosenbrock method*
ode23t	*Ordinary differential and differential algebraic equations by the trapezoidal rule*
ode23tb	*Ordinary differential equations using TR-BDF2*

The common syntax for the previous seven commands is the following:

```
[T, y] = solver(odefun,tspan,y0)
[T, y] = solver(odefun,tspan,y0,options)
[T, y] = solver(odefun,tspan,y0,options,p1,p2...)
[T, y, TE, YE, IE] = solver(odefun,tspan,y0,options)
```

In the above, solver can be any of the commands ode45, ode23, ode113, ode15s, ode23s, ode23t , or ode23tb.

The argument *odefun* evaluates the right-hand side of the differential equation or system written in the form $y' = f(t, y)$ or $M(t, y)y' = f(t, y)$, where $M(t, y)$ is called a mass matrix. The command *ode23s* can only solve equations with constant mass matrix. The commands *ode15s* and *ode23t* can solve algebraic differential equations and systems of ordinary differential equations with a singular mass matrix. The argument *tspan* is a vector that specifies the range of integration $[t_0, t_f]$ ($tspan = [t_0, t_1,...,t_f]$, which must be either an increasing or decreasing list, is used to obtain solutions for specific values of t). The argument y_0 specifies a vector of initial conditions. The arguments $p1, p2,...$ are optional parameters that are passed to *odefun*. The argument *options* specifies additional integration options using the command options *odeset* which can be found in the program manual. The vectors T and y present the numerical values of the independent and dependent variables for the solutions found.

As a first example we find solutions in the interval $[0,12]$ of the following system of ordinary differential equations:

$$y'_1 = y_2 y_3 \qquad y_1(0) = 0$$
$$y'_2 = -y_1 y_3 \qquad y_2(0) = 1$$
$$y'_3 = -0.51 y_1 y_2 \qquad y_3(0) = 1$$

For this, we define a function named *system1* in an M-file, which will store the equations of the system. The function begins by defining a column vector with three rows which are subsequently assigned components that make up the syntax of the three equations (Figure 4-28).

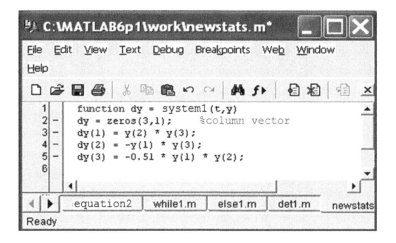

Figure 4-28.

We then solve the system by typing the following in the Command Window:

```
>> [T, Y] = ode45(@system1,[0 12],[0 1 1])
```

```
T =

0
0.0001
0.0001
0.0002
0.0002
0.0005
.
.
11.6136
11.7424
11.8712
12.0000

Y =
0 1.0000 1.0000
0.0001 1.0000 1.0000
0.0001 1.0000 1.0000
0.0002 1.0000 1.0000
0.0002 1.0000 1.0000
0.0005 1.0000 1.0000
0.0007 1.0000 1.0000
0.0010 1.0000 1.0000
0.0012 1.0000 1.0000
0.0025 1.0000 1.0000
0.0037 1.0000 1.0000
0.0050 1.0000 1.0000
0.0062 1.0000 1.0000
```

```
0.0125  0.9999  1.0000
0.0188  0.9998  0.9999
0.0251  0.9997  0.9998
0.0313  0.9995  0.9997
0.0627  0.9980  0.9990
 .
 .
 .
 0.8594-0.5105  0.7894
 0.7257-0.6876  0.8552
 0.5228-0.8524  0.9281
 0.2695-0.9631  0.9815
-0.0118-0.9990  0.9992
-0.2936-0.9540  0.9763
-0.4098-0.9102  0.9548
-0.5169-0.8539  0.9279
-0.6135-0.7874  0.8974
-0.6987-0.7128  0.8650
```

To better interpret the results, the above numerical solution can be graphed (Figure 4-29) by using the following command:

```
>> plot (T, Y(:,1), '-', T, Y(:,2),'-', T, Y(:,3),'. ')
```

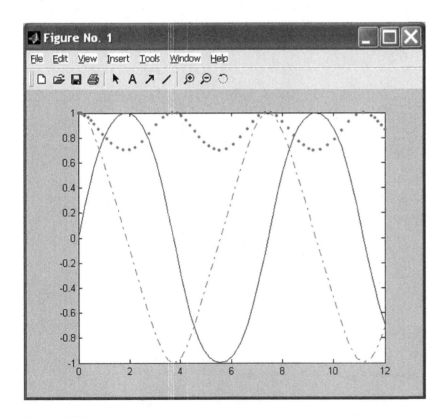

Figure 4-29.

Ordinary Differential Equations with Boundary Conditions

MATLAB also allows you to solve ordinary differential equations with boundary conditions. The boundary conditions specify a relationship that must hold between the values of the solution function at the end points of the interval on which it is defined. The simplest problem of this type is the system of equations

$$y' = f(x,y)$$

where x is the independent variable, y is the dependent variable and y' is the derivative *with respect to x* (i.e., $y' = dy/dx$). In addition, the solution on the interval $[a, b]$ has to meet the following boundary condition:

$$g(y(a), y(b)) = 0$$

More generally this type of differential equation can be expressed as follows:

$$y' = f(x,y,P)$$
$$g(y(a), y(b), P) = 0$$

where the vector p consists of parameters which have to be determined simultaneously with the solution via the boundary conditions.

The command that solves these problems is *bvp4c*, whose syntax is as follows:

```
Sol = bvp4c (odefun, bcfun, solinit)
Sol = bvp4c (odefun, bcfun, solinit, options)
Sol = bvp4c(odefun,bcfun,solinit,options,p1,p2...)
```

In the syntax above *odefun* is a function that evaluates $f(x, y)$. It may take one of the following forms:

```
dydx = odefun(x,y)
dydx = odefun(x,y,p1,p2,...)
dydx = odefun (x, y, parameters)
dydx = odefun(x,y,parameters,p1,p2,...)
```

The argument *bcfun* in *Bvp4c* is a function that computes the residual in the boundary conditions. Its form is as follows:

```
Res = bcfun (ya, yb)
Res = bcfun(ya,yb,p1,p2,...)
Res = bcfun (ya, yb, parameters)
Res = bcfun(ya,yb,parameters,p1,p2,...)
```

The argument *solinit* is a structure containing an initial guess of the solution. It has the following fields: x (which gives the ordered nodes of the initial mesh so that the boundary conditions are imposed at $a = $ solinit.x(1) and $b = $ solinit.x(end); and y (the initial guess for the solution, given as a vector, so that the i-th entry is a constant guess for the i-th component of the solution at all the mesh points given by x)) The structure *solinit* is created using the command *bvpinit*. The syntax is solinit = bvpinit(x,y).

As an example we solve the second order differential equation:

$$y'' + |y| = 0$$

whose solutions must satisfy the boundary conditions:

$$y(0) = 0$$
$$y(4) = -2$$

This is equivalent to the following problem (where $y_1 = y$ and $y_2 = y'$):

$$y_1' = y_2$$
$$y_2' = -|y_1|$$

We consider a mesh of five equally spaced points in the interval [0,4] and our initial guess for the solution is $y_1 = 1$ and $y_2 = 0$. These assumptions are included in the following syntax:

```
>> solinit = bvpinit (linspace (0,4,5), [1 0]);
```

The M-files depicted in Figures 4-30 and 4-31 show how to enter the equation and its boundary conditions.

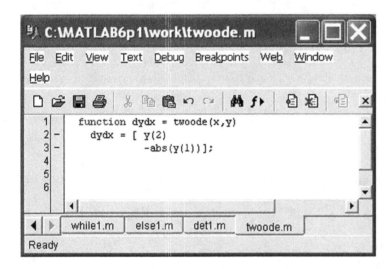

Figure 4-30.

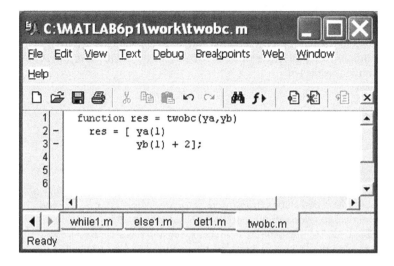

Figure 4-31.

The following syntax is used to find the solution of the equation:

```
>> Sun = bvp4c (@twoode, @twobc, solinit);
```

The solution can be graphed (Figure 4-32) using the command *bvpval* as follows:

```
>> y = bvpval (Sun, linspace (0,4));
>> plot (x, y(1,:));
```

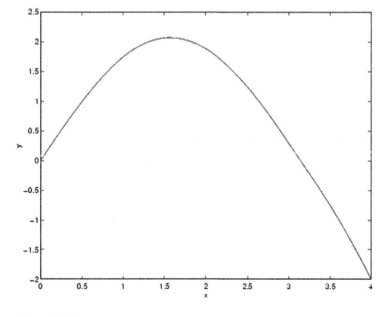

Figure 4-32.

Partial Differential Equations

MATLAB's Basic module has features that enable you to solve partial differential equations and systems of partial differential equations with initial boundary conditions. The basic function used to calculate the solutions is *pedepe*, and the basic function used to evaluate these solutions is *pdeval*.

The syntax of the function *pedepe* is as follows:

```
Sol = pdepe (m, pdefun, icfun, bcfun, xmesh, tspan)
Sol = pdepe (m, pdefun, icfun, bcfun, xmesh, tspan, options)
Sun= pdepe(m,pdefun,icfun,bcfun,xmesh,tspan,options,p1,p2...)
```

The parameter *m* takes the value 0, 1 or 2 according to the nature of the symmetry of the problem (block, cylindrical or spherical, respectively). The argument *pdefun* defines the components of the differential equation, *icfun* defines the initial conditions, *bcfun* defines the boundary conditions, *xmesh* and *tspan* are vectors $[x_0, x_1, ..., x_n]$ and $[t_0, t_1, ... , t_f]$ that specify the points at which a numerical solution is requested ($n, f \geq 3$), *options* specifies some calculation options of the underlying solver (RelTol, AbsTol, NormControl, InitialStep and MaxStep to specify relative tolerance, absolute tolerance, norm tolerance, initial step and max step, respectively) and *p1*, *p2*,... are parameters to pass to the functions *pdefun*, *icfun* and *bcfun*.

pdepe solves partial differential equations of the form:

$$c\left(x,t,u,\frac{\partial u}{\partial x}\right)\frac{\partial u}{\partial t} = x^{-m}\frac{\partial}{\partial x}\left(x^m f\left(x,t,u,\frac{\partial u}{\partial x}\right)\right)+ s\left(x,t,u,\frac{\partial u}{\partial x}\right)$$

Where $a \leq x \leq b$ and $t_0 \leq t \leq t_f$. Moreover, for $t = t_0$ and for all x the solution components meet the initial conditions:

$$u(x,t_0) = u_0(x)$$

and for all t and each $x = a$ or $x = b$, the solution components satisfy the boundary conditions of the form:

$$p(x,t,u) + q(x,t)f\left(x,t,u,\frac{\partial u}{\partial x}\right) = 0$$

In addition, we have that a = xmesh (1), b = xmesh (end), tspan (1) $=t_0$ and tspan (end) = t_f. Moreover *pdefun* finds the terms *c*, *f* and *s* of the partial differential equation, so that:

```
[c, f, s] = pdefun (x, t, u, dudx)
```

Similiarly *icfun* evaluates the initial conditions

```
u = icfun (x)
```

Finally, *bcfun* evaluates the terms *p* and *q* of the boundary conditions:

```
[pl, ql, pr, qr] = bcfun (xl, ul, xr, ur, t)
```

As a first example we solve the following partial differential equation ($x \in [0,1]$ and $t \geq 0$):

$$\pi^2 \frac{\partial u}{\partial t} = \frac{\partial}{\partial x}\left(\frac{\partial u}{\partial x}\right)$$

satisfying the initial condition:

$$u(x,0) = \sin \pi x$$

and the boundary conditions:

$$u(0,t) \equiv 0$$

$$\pi e^{-t} + \frac{\partial u}{\partial x}(1,t) = 0$$

We begin by defining functions in M-files as shown in Figures 4-33 to 4-35.

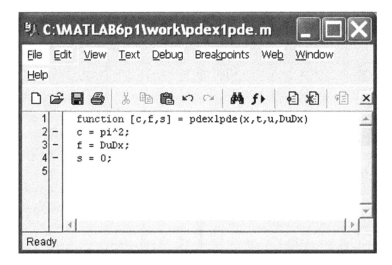

Figure 4-33.

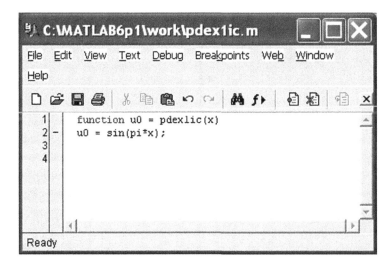

Figure 4-34.

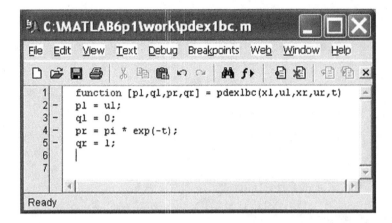

```
function [pl,ql,pr,qr] = pdexlbc(xl,ul,xr,ur,t)
pl = ul;
ql = 0;
pr = pi * exp(-t);
qr = l;
```

Figure 4-35.

Once the support functions have been defined, we define the function that solves the equation (see the M-file in Figure 4-36).

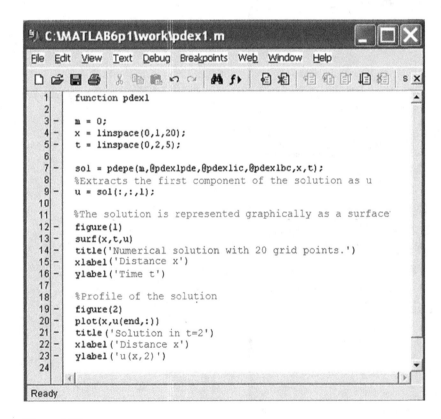

```
function pdex1

m = 0;
x = linspace(0,1,20);
t = linspace(0,2,5);

sol = pdepe(m,@pdexlpde,@pdexlic,@pdexlbc,x,t);
%Extracts the first component of the solution as u
u = sol(:,:,1);

%The solution is represented graphically as a surface
figure(1)
surf(x,t,u)
title('Numerical solution with 20 grid points.')
xlabel('Distance x')
ylabel('Time t')

%Profile of the solution
figure(2)
plot(x,u(end,:))
title('Solution in t=2')
xlabel('Distance x')
ylabel('u(x,2)')
```

Figure 4-36.

To view the solution (Figures 4-37 and 4-38), we enter the following into the MATLAB Command Window:

```
>> pdex1
```

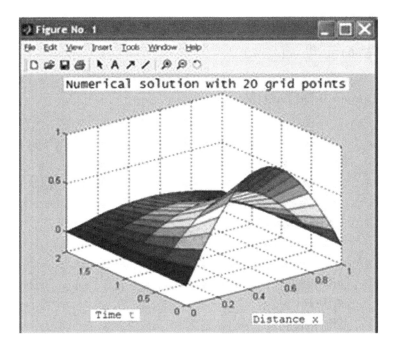

Figure 4-37.

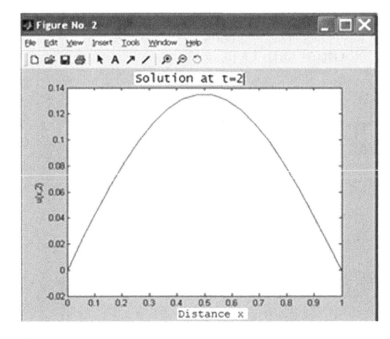

Figure 4-38.

As a second example we solve the following system of partial differential equations ($x \in [0,1]$ and $t \geq 0$):

$$\frac{\partial u_1}{\partial t} = 0.024 \frac{\partial^2 u_1}{\partial x^2} - F(u_1 - u_2)$$

$$\frac{\partial u_2}{\partial t} = 0.170 \frac{\partial^2 u_2}{\partial x^2} - F(u_1 - u_2)$$

$$F(y) = \exp(5.73y) - \exp(-11.46y)$$

satisfying the initial conditions:

$$u_1(x,0) \equiv 1$$
$$u_2(x,0) \equiv 0$$

and the boundary conditions:

$$\frac{\partial u_1}{\partial x}(0,t) \equiv 0$$
$$u_2(0,t) \equiv 0$$
$$u_1(1,t) \equiv 1$$
$$\frac{\partial u_2}{\partial x}(1,t) \equiv 0$$

To conveniently use the function *pdepe*, the system can be written as:

$$\begin{bmatrix} 1 \\ 1 \end{bmatrix} .* \frac{\partial}{\partial t} \begin{bmatrix} u_1 \\ u_2 \end{bmatrix} = \frac{\partial}{\partial x} \begin{bmatrix} 0.024(\partial u_1 / \partial x) \\ 0.170(\partial u_2 / \partial x) \end{bmatrix} + \begin{bmatrix} -F(u_1 - u_2) \\ F(u_1 - u_2) \end{bmatrix}$$

The left boundary condition can be written as:

$$\begin{bmatrix} 0 \\ u_2 \end{bmatrix} + \begin{bmatrix} 1 \\ 0 \end{bmatrix} .* \begin{bmatrix} 0.024(\partial u_1 / \partial x) \\ 0.170(\partial u_2 / \partial x) \end{bmatrix} = \begin{bmatrix} 0 \\ 0 \end{bmatrix}$$

and the right boundary condition can be written as:

$$\begin{bmatrix} u_1 - 1 \\ 0 \end{bmatrix} + \begin{bmatrix} 1 \\ 0 \end{bmatrix} .* \begin{bmatrix} 0.024(\partial u_1 / \partial x) \\ 0.170(\partial u_2 / \partial x) \end{bmatrix} = \begin{bmatrix} 0 \\ 0 \end{bmatrix}$$

We start by defining the functions in M-files as shown in Figures 4-39 to 4-41.

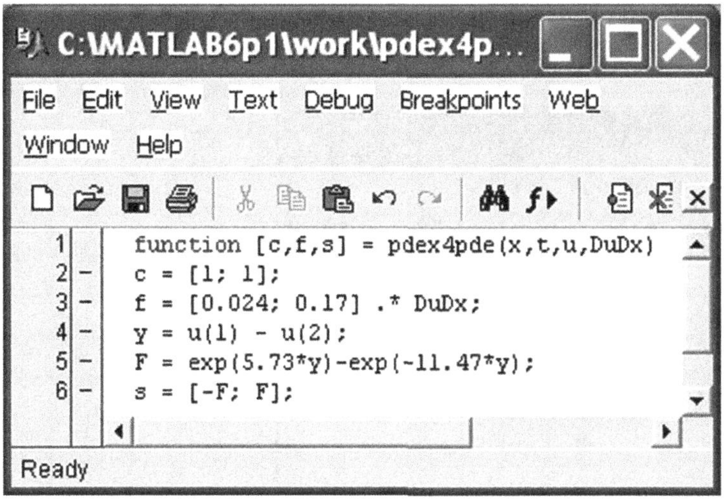

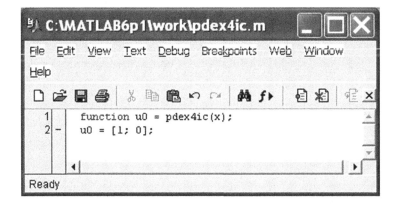

```
     function [c,f,s] = pdex4pde(x,t,u,DuDx)
1
2 -  c = [1; 1];
3 -  f = [0.024; 0.17] .* DuDx;
4 -  y = u(1) - u(2);
5 -  F = exp(5.73*y)-exp(-11.47*y);
6 -  s = [-F; F];
```

Figure 4-39.

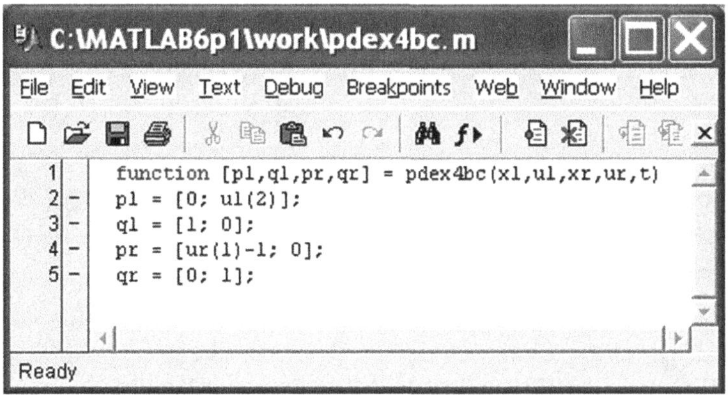

```
     function [pl,ql,pr,qr] = pdex4bc(xl,ul,xr,ur,t)
1
2 -  pl = [0; ul(2)];
3 -  ql = [1; 0];
4 -  pr = [ur(1)-1; 0];
5 -  qr = [0; 1];
```

Figure 4-40.

```
     function u0 = pdex4ic(x);
1
2 -  u0 = [1; 0];
```

Figure 4-41.

Once the support functions are defined, the function that solves the system of equations is given by the M-file shown in Figure 4-42.

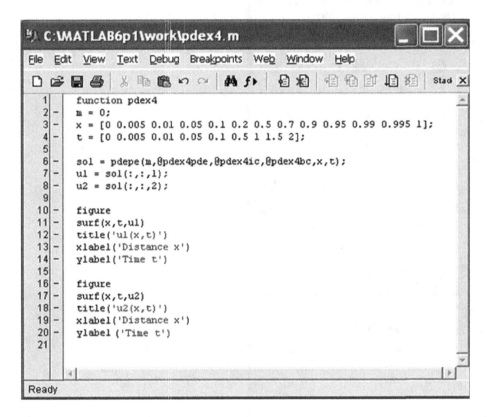

```
function pdex4
m = 0;
x = [0 0.005 0.01 0.05 0.1 0.2 0.5 0.7 0.9 0.95 0.99 0.995 1];
t = [0 0.005 0.01 0.05 0.1 0.5 1 1.5 2];

sol = pdepe(m,@pdex4pde,@pdex4ic,@pdex4bc,x,t);
u1 = sol(:,:,1);
u2 = sol(:,:,2);

figure
surf(x,t,u1)
title('u1(x,t)')
xlabel('Distance x')
ylabel('Time t')

figure
surf(x,t,u2)
title('u2(x,t)')
xlabel('Distance x')
ylabel ('Time t')
```

Figure 4-42.

To view the solution (Figures 4-43 and 4-44), we enter the following in the MATLAB Command Window:

>> pdex4

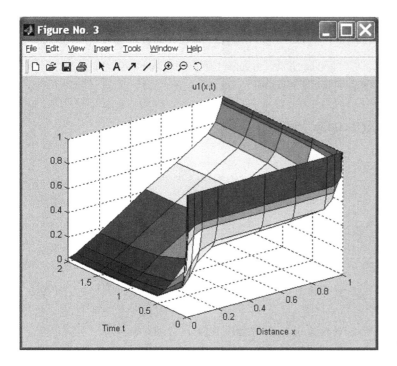

Figure 4-43.

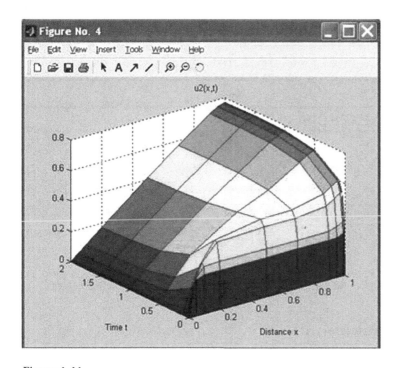

Figure 4-44.

EXERCISE 4-1

Minimize the function $x^3 - 2x - 5$ in the interval (0,2) and calculate the value that the function takes at that point, displaying information about all iterations of the optimization process.

```
>> f = inline('x.^3-2*x-5');
>> [x,fval] = fminbnd(f, 0, 2,optimset('Display','iter'))
```

Func-count	x	f(x)	Procedure
1	0.763932	-6.08204	initial
2	1.23607	-5.58359	golden
3	0.472136	-5.83903	golden
4	0.786475	-6.08648	parabolic
5	0.823917	-6.08853	parabolic
6	0.8167	-6.08866	parabolic
7	0.81645	-6.08866	parabolic
8	0.816497	-6.08866	parabolic
9	0.81653	-6.08866	parabolic

```
Optimization terminated successfully:
the current x satisfies the termination criteria using OPTIONS.TolX of 1.000000e-004

x =

0.8165

fval =

-6.0887
```

EXERCISE 4-2

Find in a neighborhood of $x = 1.3$ a zero of the function:

$$f(x) = \frac{1}{(x-0.3)^2 + 0.01} + \frac{1}{(x-0.9)^2 + 0.04} - 6.$$

Minimize this function on the interval (0,2).

First we find a zero of the function using the initial estimate of $x = 1.3$, presenting information about the iterations and checking that the result is indeed a zero.

```
>> [x,feval]=fzero(inline('1/((x-0.3)^2+0.01)+...
```

```
1/((x-0.9)^2+0.04)-6'),1.3,optimset('Display','iter'))
Func-count       x        f(x)  Procedure
1              1.3  -0.00990099    initial
2          1.26323    0.882416     search

Looking for a zero in the interval [1.2632, 1.3]

3          1.29959   -0.00093168       interpolation
4          1.29955  1.23235e-007       interpolation
5          1.29955 -1.37597e-011       interpolation
6          1.29955            0        interpolation
Zero found in the interval: [1.2632, 1.3].

x =

1.2995

feval =

0
```

Secondly, we minimize the function specified in the interval [0,2] and also present information about the iterative process, terminating the process when the value of x which minimizes the function is found. In addition, the value of the function at this point is calculated.

```
>> [x,feval]=fminbnd(inline('1/((x-0.3)^2+0.01)+...
1/((x-0.9)^2+0.04)-6'),0,2,optimset('Display','iter'))

Func-count        x        f(x)  Procedure
1          0.763932    15.5296      initial
2           1.23607    1.66682      golden
3           1.52786   -3.03807      golden
4            1.8472   -4.51698    parabolic
5           1.81067   -4.41339    parabolic
6           1.90557   -4.66225      golden
7           1.94164   -4.74143      golden
8           1.96393   -4.78683      golden
9           1.97771   -4.81365      golden
10          1.98622   -4.82978      golden
11          1.99148   -4.83958      golden
12          1.99474   -4.84557      golden
13          1.99675   -4.84925      golden
14          1.99799   -4.85152      golden
15          1.99876   -4.85292      golden
16          1.99923   -4.85378      golden
17          1.99953   -4.85431      golden
18          1.99971   -4.85464      golden
19          1.99982   -4.85484      golden
20          1.99989   -4.85497      golden
21          1.99993   -4.85505      golden
22          1.99996   -4.85511      golden
```

173

```
Optimization terminated successfully:
the current x satisfies the termination criteria using OPTIONS.TolX of 1.000000e-004

x =

2.0000

feval =

-4.8551
```

EXERCISE 4-3

The intermediate value theorem says that if f is a continuous function on the interval [a, b] and L is a number between f(a) and f(b), then there is a c (a< c < b) such that f(c) = L. For the function f(x) = cos(x−1), find the value c in the interval [1, 2.5] such that f(c)= 0.8.

The question asks us to solve the equation cos(x− 1) − 0.8 = 0 in the interval [1, 2.5].

```
>> c = fzero (inline ('cos (x-1) - 0.8'), [1 2.5])

c =

1.6435
```

EXERCISE 4-4

Calculate the following integral using both Simpson's and Lobato's methods:

$$\int_1^6 \left(2+\sin\left(2\sqrt{x}\right)dx\cdot\right)$$

For the solution using Simpson's method we have:

```
>> quad(inline('2+sin(2*sqrt(x))'),1,6)

ans =

8.1835
```

For the solution using Lobato's method we have:

```
>> quadl(inline('2+sin(2*sqrt(x))'),1,6)
```

ans =

8.1835

EXERCISE 4-5

Calculate the area under the normal curve (0,1) between the limits−1.96 and 1.96.

The integral we need to calculate is $\int_{-196}^{196} \frac{e^{\frac{-x^2}{2}}}{\sqrt{2\pi}} dx$.

The calculation is done in MATLAB using Lobato's method as follows:

```
(((>> quadl(inline('exp(-x.^2/2)/sqrt(2*pi)'), - 1.96,1.96)
```

ans =

0.9500

EXERCISE 4-6

Calculate the volume of the hemisphere-function defined in

$$[-1,1]\times[-1,1] by f(x,y) = \sqrt{1-\left(x^2+y^2\right)}$$

```
>> dblquad(inline('sqrt(max(1-(x.^2+y.^2),0))'),-1,1,-1,1)
```

ans =

2.0944

The calculation could also have been done in the following way:

```
>> dblquad(inline('sqrt(1-(x.^2+y.^2)).*(x.^2+y.^2<=1)'),-1,1,-1,1)
```

ans =

2.0944

EXERCISE 4-7

Evaluate the following double integral:

$$\int_{3}^{4}\int_{1}^{2}\frac{1}{(x+y)^{2}}\,dxdy\cdot$$

`(>> dblquad(inline('1./(x+y).^2'),3,4,1,2)`

ans =

0.0408

EXERCISE 4-8

Solve the following Van der Pol system of equations:

$$y'_{1}=y_{2} \qquad\qquad y_{1}(0)=0$$
$$y'_{2}=1000\left(1-y^{2}_{1}\right)y_{2}-y_{1}\quad y_{2}(0)=1$$

We begin by defining a function named *vdp100* in an M-file, where we will store the equations of the system. This function begins by defining a vector column with two empty rows which are subsequently assigned the components which make up the equation (Figure 4-45).

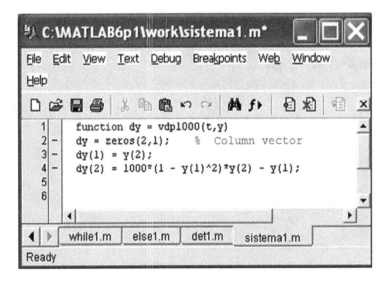

Figure 4-45.

We then solve the system and plot the solution $y_1 = y_1(t)$ given by the first column (Figure 4-46) by typing the following into the Command Window:

```
>> [T, Y] = ode15s(@vdp1000,[0 3000],[2 0]);
>> plot (T, Y(:,1),'-')
```

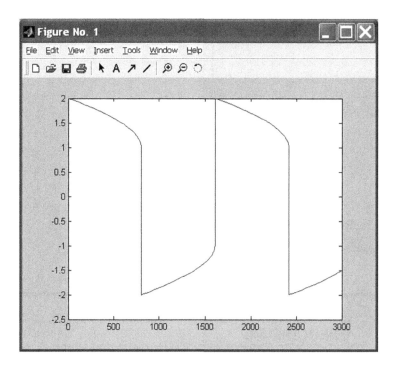

Figure 4-46.

Similarly we plot the solution $y_2 = y_2(t)$ (Figure 4-47) by using the syntax:

```
>> plot (T, Y(:,2),'-')
```

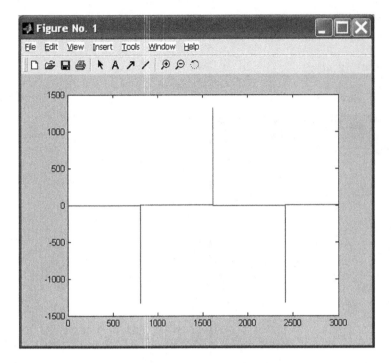

Figure 4-47.

EXERCISE 4-9

Given the following differential equation

$$y'' + (\lambda - 2q\cos(2x))y = 0$$

subject to the boundary conditions y(0) = 1, y'(0) = 0, y'(π) = 0, find a solution for q = 5 and λ = 15 based on an initial solution defined on 10 equally spaced points in the interval [0, π] and graph the first component of the solution on 100 equally spaced points in the interval [0, π].

The given equation is equivalent to the following system of first order differential equations:

$$y'_1 = y_2$$
$$y_2' = -(\lambda - 2q\cos 2x)y_1$$

with the following boundary conditions:

$$y_1(0) - 1 = 0$$
$$y_2(0) = 0$$
$$y_2(\pi) = 0$$

The system of equations is introduced in the M-file shown in Figure 4-48, the boundary conditions are given in the M-file shown in Figure 4-49, and the M-file in Figure 4-50 sets up the initial solution.

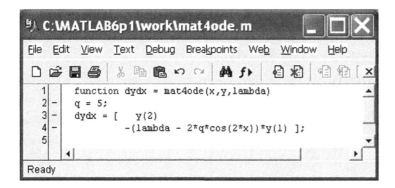

Figure 4-48.

```
C:\MATLAB6p1\work\mat4bc.m
File  Edit  View  Text  Debug  Breakpoints  Web  Window  Help

1    function res = mat4bc(ya,yb,lambda)
2 -  res = [  ya(2)
3 -           yb(2)
4 -           ya(1)-1 ];
5
Ready
```

Figure 4-49.

```
C:\MATLAB6p1\work\mat4init.m
File  Edit  View  Text  Debug  Breakpoints  Web  Window  Help

1    function yinit = mat4init(x)
2 -  yinit = [  cos(4*x)
3 -            -4*sin(4*x) ];
4
5
Ready
```

Figure 4-50.

The initial solution for $\lambda = 15$ and 10 equally spaced points in $[0, \pi]$ is calculated using the following MATLAB syntax:

```
>> lambda = 15;
solinit = bvpinit (linspace(0,pi,10), @mat4init, lambda);
```

The numerical solution of the system is calculated using the following syntax:

```
>> sol = bvp4c(@mat4ode,@mat4bc,solinit);
```

To graph the first component on 100 equally spaced points in the interval $[0, \pi]$ we use the following syntax:

```
>> xint = linspace(0,pi);
Sxint = bvpval (ground, xint);
plot (xint, Sxint(1,:)))
axis([0 pi-1 1.1])
xlabel ('x')
ylabel('solution y')
```

The result is shown in Figure 4-51.

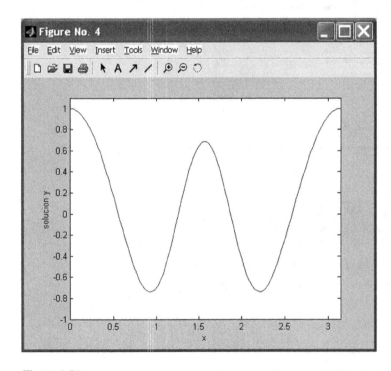

Figure 4-51.

EXERCISE 4-10

Solve the following differential equation

$$y'' + (1 - y^2)y' + y = 0$$

in the interval [0,20], taking as initial solution y = 2, y' = 0. Solve the more general equation

$$y'' + \mu(1 - y^2)y' + y = 0 \quad \mu > 0 \cdot$$

The general equation above is equivalent to the following system of first-order linear equations:

$$y'_1 = y_2$$
$$y'_2 = \mu(1 - y_1^2)y_2 - y_1$$

which is defined for $\mu = 1$ in the M-file shown in Figure 4-52.

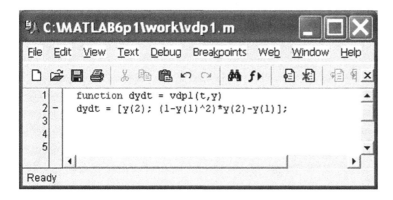

Figure 4-52.

Taking the initial solution $y_1 = 2$ and $y_2 = 0$ in the interval [0,20], we can solve the system using the following MATLAB syntax:

```
>> [t, y] = ode45(@vdp1,[0 20],[2; 0])

t =

0
0.0000
0.0001
0.0001
0.0001
0.0002
```

```
0.0004
0.0005
0.0006
0.0012
.
.
.
19.9559
19.9780
20.0000


y =

2.0000  0
2.0000 - 0.0001
2.0000 - 0.0001
2.0000 - 0.0002
2.0000 - 0.0002
2.0000 - 0.0005
.
.
.
1.8729 1.0366
1.9358 0.7357
1.9787 0.4746
2.0046 0.2562
2.0096 0.1969
2.0133 0.1413
2.0158 0.0892
2.0172 0.0404
```

We can graph the solutions using the following syntax (see Figure 4-53):

```
>> plot (t, y(:,1),'-', t, y(:,2),'-')
>> xlabel ('time t')
>> ylabel('solution y')
>> legend ('y_1', 'y_2')
```

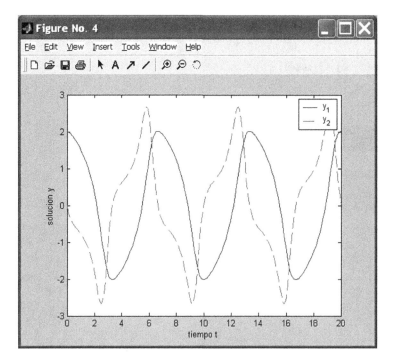

Figure 4-53.

To solve the general system with the parameter μ, we define the system in the M-file shown in Figure 4-54.

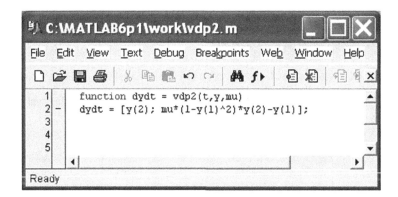

Figure 4-54.

Now we can graph the first solution $y_1 = 2$ and $y_2 = 0$ corresponding to $\mu = 1000$ in the interval [0,1500] using the following syntax (see Figure 4-55):

```
>> [t, y] = ode15s(@vdp2,[0 1500],[2; 0],[],1000);
>> xlabel ('time t')
>> ylabel ('solution y_1')
```

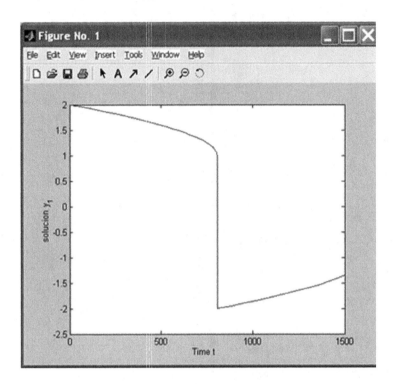

Figure 4-55.

To graph the first solution $y_1 = 2$ and $y_2 = 0$ for another value of the parameter, for example $\mu = 100$, in the interval [0,1500], we use the following syntax (see Figure 4-56):

```
>> [t, y] = ode15s(@vdp2,[0 1500],[2; 0],[],100);
>> plot (t, y(:,1),'-');
```

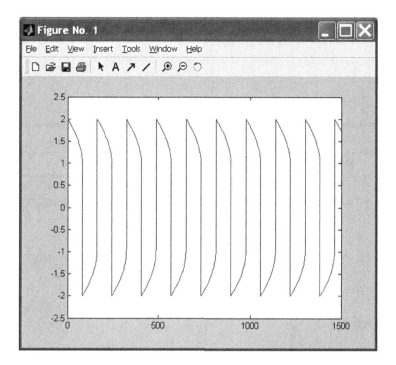

Figure 4-56.

EXERCISE 4-11

The Fibonacci sequence {an} is defined by the recurrence law $a_1 = 1$, $a_2 = 1$, $a_{n+1} = a_{n-1} + a_n$. Represent this sequence by a recursive function and calculate a_2, a_5 and a_{20}.

To generate terms of the Fibonacci sequence we define a recursive function in the M-file *fibo.m* shown in Figure 4-57.

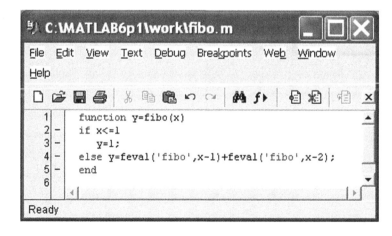

Figure 4-57.

Terms 2, 5 and 20 of the sequence are now calculated using the syntax:

```
>> [fibo (2), fibo (5), fibo (20)]

ans =

    2  8  10946
```

EXERCISE 4-12

Define the Kronecker delta, which equals 1 if $x = 0$ and 0 otherwise. Define the modified Kronecker delta function, which is 0 if $x = 0$, 1 if $x > 0$ and -1 if $x < 0$ and graph it. Lastly, define the piecewise function that is equal to 0 if $x \leq -3$, x^3 if $-3 < x < -2$, x^2 if $-2 \leq x \leq 2$, x if $2 < x < 3$ and 0 if $3 \leq x$, and graph it.

The Kronecker delta *delta(x)* is defined in the M-file *delta.m* shown in Figure 4-58. The modified Kronecker delta *delta1(x)* is defined in the M-file *delta1.m* shown in Figure 4-59. To define the third function *piece1(x)* of the exercise, we create the M-file *piece1.m* shown in Figure 4-60.

Figure 4-58.

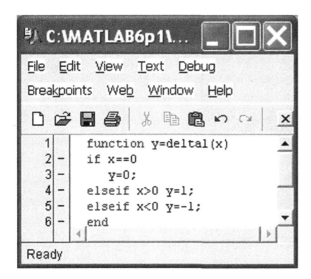

```
function y=deltal(x)
if x==0
    y=0;
elseif x>0 y=1;
elseif x<0 y=-1;
end
```

Figure 4-59.

```
function y=piece1(x)
if x<=-3
    y=0;
elseif -3<x & x<-2
    y=x^3;
elseif -2<=x & x<=2
    y=x^2;
elseif 2<x & x<3
    y=x
elseif x>=3
    y=0;
end
```

Figure 4-60.

To graphically represent the modified Kronecker delta on the domain [−10, 10] (and with codomain [− 2, 2]) we use the following syntax(see Figure 4-61):

```
>> fplot ('delta1 (x)', [- 10 10 - 2-2])
>> title 'Modified Kronecker Delta'
```

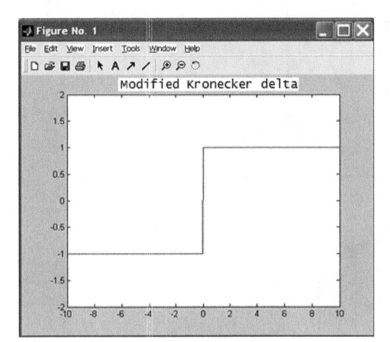

Figure 4-61.

To graphically represent the piecewise function on the interval [- 5,5] we use the following syntax (see Figure 4-62):

```
>> fplot ('piece1 (x)', [- 5 5]);
>> title 'Piecewise function'
```

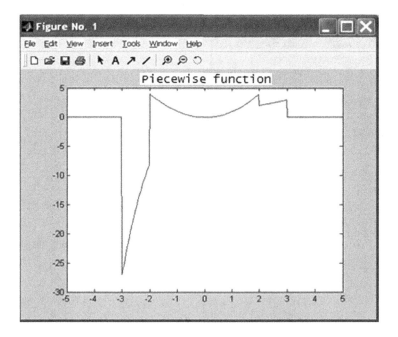

Figure 4-62.

Define a function descriptive(v) which returns the variance and coefficient of variation of the elements of a given vector v. As an application, find the variance and coefficient of variation of the set of numbers 1, 5, 6, 7 and 9.

Figure 4-63 shows the M-file which defines the function *descriptive*.

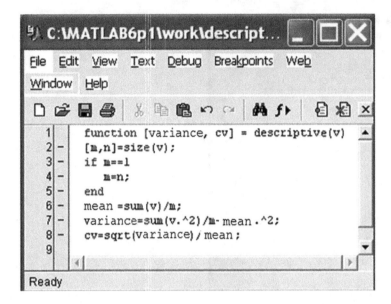

```
function [variance, cv] = descriptive(v)
[m,n]=size(v);
if m==1
    m=n;
end
mean =sum(v)/m;
variance=sum(v.^2)/m- mean .^2;
cv=sqrt(variance)/ mean ;
```

Figure 4-63.

To find the variance and coefficient of variation of the given set of numbers, we use the following syntax:

```
>> [variance, cv] = descriptive([1 5 6 7 9])

variance =

    7.0400

cv =

0.4738
```

CHAPTER 5

Numerical Algorithms: Equations, Derivatives and Integrals

Solving Non-Linear Equations

MATLAB is able to implement a number of algorithms which provide numerical solutions to certain problems which play a central role in the solution of non-linear equations. Such algorithms are easy to construct in MATLAB and are stored as M-files. From previous chapters we know that an M-file is simply a sequence of MATLAB commands or functions that accept arguments and produces output. The M-files are created using the text editor.

The Fixed Point Method for Solving x = g (x)

The fixed point method solves the equation $x = g(x)$, under certain conditions on the function g, using an iterative method that begins with an initial value p_0 (a first approximation to the solution) and defines $p_{k+1} = g(p_k)$. The fixed point theorem ensures that, in certain circumstances, this sequence will converges to a solution of the equation $x = g(x)$. In practice the iterative process will stop when the absolute or relative error corresponding to two consecutive iterations is less than a preset value (*tolerance*). The smaller this value, the better the approximation to the solution of the equation.

This simple iterative method can be implemented using the M-file shown in Figure 5-1.

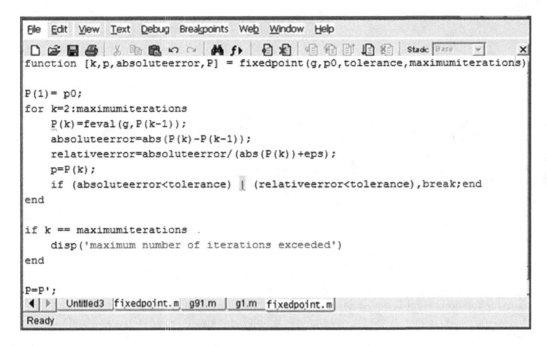

```
File  Edit  View  Text  Debug  Breakpoints  Web  Window  Help

function [k,p,absoluteerror,P] = fixedpoint(g,p0,tolerance,maximumiterations)

P(1)= p0;
for k=2:maximumiterations
    P(k)=feval(g,P(k-1));
    absoluteerror=abs(P(k)-P(k-1));
    relativeerror=absoluteerror/(abs(P(k))+eps);
    p=P(k);
    if (absoluteerror<tolerance) | (relativeerror<tolerance),break;end
end

if k == maximumiterations
    disp('maximum number of iterations exceeded')
end

P=P';
```

Figure 5-1.

As an example we solve the following non-linear equation:

$$x - 2^{-x} = 0.$$

In order to apply the fixed point algorithm we write the equation in the form $x = g(x)$ as follows:

$$x - 2^{-x} = g(x).$$

We will start by finding an approximate solution which will be the first term p_0. To plot the x axis and the curve defined by the given equation on the same graph we use the following syntax (see Figure 5-2):

```
>> fplot ('[x-2^(-x), 0]',[0, 1])
```

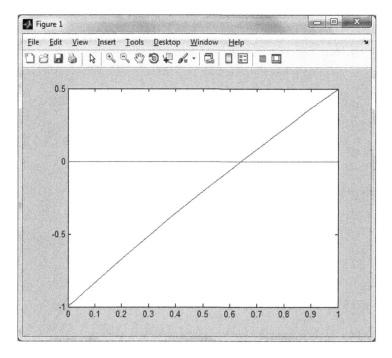

Figure 5-2.

The graph shows that one solution is close to $x = 0.6$. We can take this value as the initial value. We choose $p_0 = 0.6$. If we consider a tolerance of 0.0001 for a maximum of 1000 iterations, we can solve the problem once we have defined the function $g(x)$ in the M-file *g1.m* (see Figure 5-3).

Figure 5-3.

We can now solve the equation using the MATLAB syntax:

```
>> [k, p] = fixedpoint('g1',0.6,0.0001,1000)

k =

    10

p =

    0.6412
```

We obtain the solution $x = 0.6412$ at the 1000th iteration. To check if the solution is approximately correct, we must verify that g1(0.6412) is close to 0.6412.

```
>> g1 (0.6412)
```

ans =

    0.6412

Thus we observe that the solution is acceptable.

Newton's Method for Solving the Equation f (x) =0

Newton's method (also called the Newton–Raphson method) for solving the equation $f(x) = 0$, under certain conditions on f, uses the iteration

$$x_{r+1} = x_r - f(x_r) / f'(x_r)$$

for an initial value x_0 close to a solution.

The M-file in Figure 5-4 shows a program which solves equations by Newton's method to a given precision.

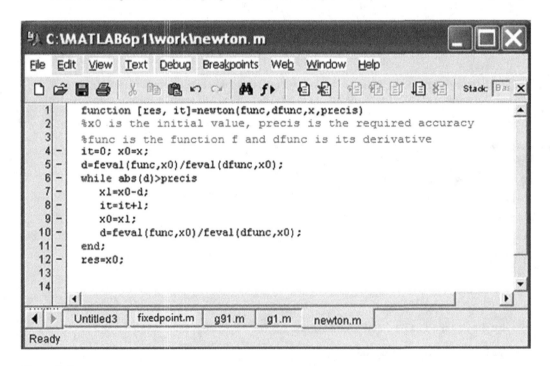

```
function [res, it]=newton(func,dfunc,x,precis)
%x0 is the initial value, precis is the required accuracy
%func is the function f and dfunc is its derivative
it=0; x0=x;
d=feval(func,x0)/feval(dfunc,x0);
while abs(d)>precis
    x1=x0-d;
    it=it+1;
    x0=x1;
    d=feval(func,x0)/feval(dfunc,x0);
end;
res=x0;
```

Figure 5-4.

As an example we solve the following equation by Newton's method:

$$x^2 - x - \sin(x + 0.15) = 0.$$

The function $f(x)$ is defined in the M-file *f1.m* (see Figure 5-5), and its derivative $f'(x)$ is given in the M-file *derf1.m* (see Figure 5-6).

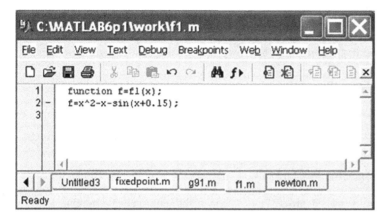

Figure 5-5.

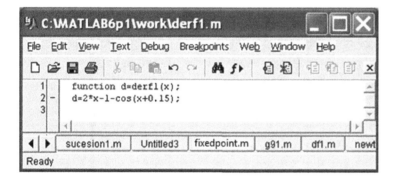

Figure 5-6.

We can now solve the equation up to an accuracy of 0.0001 and 0.000001 using the following MATLAB syntax, starting with an initial estimate of 1.5:

```
>> [x,it]=newton('f1','derf1',1.5,0.0001)

x =

1.6101

it =

2

>> [x,it]=newton('f1','derf1',1.5,0.000001)
x =
```

1.6100

it =

3

Thus we have obtained the solution $x = 1.61$ in just 2 iterations for a precision of 0.0001 and in just 3 iterations for a precision of 0.000001.

Schröder's Method for Solving the Equation f (x) =0

Schröder's method, which is similar to Newton's method, solves the equation $f(x) = 0$, under certain conditions on f, via the iteration

$$X_{r+1} = X_r - mf(X_r)/f'(X_r)$$

for an initial value x_0 close to a solution, and where m is the order of multiplicity of the solution being sought.

The M-file shown in Figure 5-7 gives the function that solves equations by Schröder's method to a given precision.

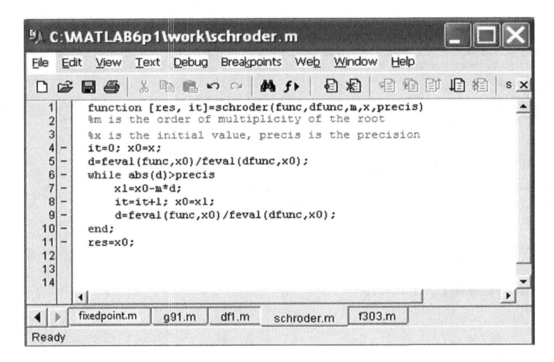

```
function [res, it]=schroder(func,dfunc,m,x,precis)
%m is the order of multiplicity of the root
%x is the initial value, precis is the precision
it=0; x0=x;
d=feval(func,x0)/feval(dfunc,x0);
while abs(d)>precis
    x1=x0-m*d;
    it=it+1; x0=x1;
    d=feval(func,x0)/feval(dfunc,x0);
end;
res=x0;
```

Figure 5-7.

Systems of Non-Linear Equations

As for differential equations, it is possible to implement algorithms with MATLAB that solve systems of non-linear equations using classical iterative numerical methods.

Among a diverse collection of existing methods we will consider the Seidel and Newton–Raphson methods.

The Seidel Method

The Seidel method for solving systems of equations is a generalization of the fixed point iterative method for single equations.

In the case of a system of two equations $x=g_1(x,y)$ and $y=g_2(x,y)$ the terms of the iteration are defined as:

$P_{k+1}=g_1(p_k,q_k)$ and $q_{k+1}=g_2(p_k,q_k)$.

Similarly, in the case of a system of three equations $x=g_1(x,y,z)$,

$y=g_2(x,y,z)$ and $z=g_3(x,y,z)$ the terms of the iteration are defined as:

$p_{k+1}=g_1(p_k,q_k,r_k)$, $q_{k+1}=g_2(p_k,q_k,r_k)$ and $r_{k+1}=g_3(p_k,q_k,r_k)$.

The M-file shown in Figure 5-8 gives a function which solves systems of equations using Seidel's method up to a specified accuracy.

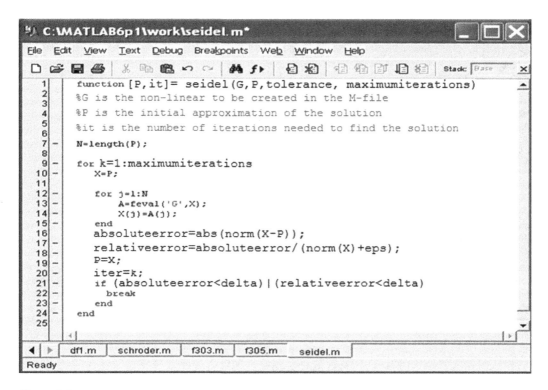

Figure 5-8.

The Newton–Raphson Method

The Newton–Raphson method for solving systems of equations is a generalization of Newton's method for single equations.

The idea behind the algorithm is familiar. The solution of the system of non-linear equations $F(X) = 0$ is obtained by generating from an initial approximation P_0 a sequence of approximations P_k which converges to the solution. Figure 5-9 shows the M-file containing the function which solves systems of equations using the Newton–Raphson method up to a specified degree of accuracy.

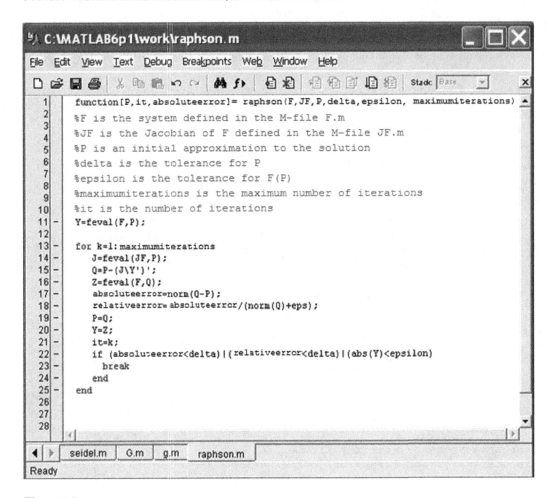

```
C:\MATLAB6p1\work\raphson.m

File  Edit  View  Text  Debug  Breakpoints  Web  Window  Help

 1    function[P,it,absoluteerror]= raphson(F,JF,P,delta,epsilon, maximumiterations)
 2    %F is the system defined in the M-file F.m
 3    %JF is the Jacobian of F defined in the M-file JF.m
 4    %P is an initial approximation to the solution
 5    %delta is the tolerance for P
 6    %epsilon is the tolerance for F(P)
 7    %maximumiterations is the maximum number of iterations
 8    %it is the number of iterations
 9
10    Y=feval(F,P);
11
12
13    for k=1:maximumiterations
14        J=feval(JF,P);
15        Q=P-(J\Y')';
16        Z=feval(F,Q);
17        absoluteerror=norm(Q-P);
18        relativeerror= absoluteerror/(norm(Q)+eps);
19        P=Q;
20        Y=Z;
21        it=k;
22        if (absoluteerror<delta)|(relativeerror<delta)|(abs(Y)<epsilon)
23           break
24        end
25    end
26
27
28

seidel.m    G.m    g.m    raphson.m

Ready
```

Figure 5-9.

As an example we solve the following system of equations by the Newton–Raphson method:

$$x^2 - 2x - y = -0.5$$
$$x^2 + 4y^2 - 4 = 0$$

taking as an initial approximation to the solution $P = [2\ 3]$.

We start by defining the system $F(X) = 0$ and its Jacobian matrix JF according to the M-files $F.m$ and $JF.m$ shown in Figures 5-10 and 5-11.

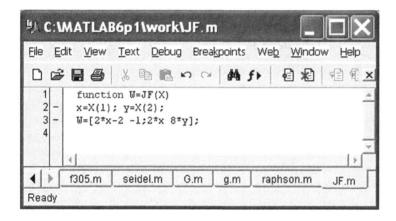

Figure 5-10.

Figure 5-11.

Then the system is solved with a tolerance of 0.00001 and with a maximum of 100 iterations using the following MATLAB syntax:

```
>> [P,it,absoluteerror]=raphson('F','JF',[2 3],0.00001,0.00001,100)
P =

1.9007 0.3112

it =

6

absoluteerror =

8. 8751e-006
```

The solution obtained in 6 iterations is $x = 1.9007$, $y = 0.3112$, with an absolute error of 8.8751e- 006.

Interpolation Methods

There are many different methods available to find an interpolating polynomial that fits a given set of points in the best possible way.

Among the most common methods of interpolation, we have Lagrange polynomial interpolation, Newton polynomial interpolation and Chebyshev approximation.

Lagrange Polynomial Interpolation

The Lagrange interpolating polynomial which passes through the $N+1$ points (x_k, y_k), $k = 0, 1, ..., N$, is defined as follows:

$$P(x) = \sum_{k=0}^{N} y_k L_{N,k}(x)$$

where:

$$L_{N,k}(x) = \frac{\displaystyle\prod_{\substack{j=0 \\ j \neq k}}^{N} (x - x_j)}{\displaystyle\prod_{\substack{j=0 \\ j \neq k}}^{N} (x_k - x_j)}.$$

The algorithm for obtaining P and L is easily implemented by the M-file shown in Figure 5-12.

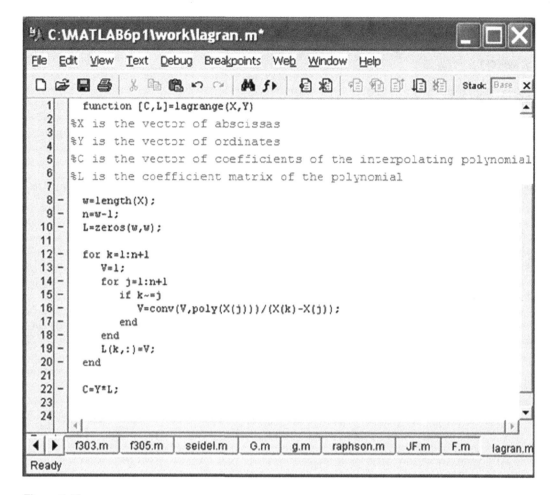

```
function [C,L]=lagrange(X,Y)
%X is the vector of abscissas
%Y is the vector of ordinates
%C is the vector of coefficients of the interpolating polynomial
%L is the coefficient matrix of the polynomial

w=length(X);
n=w-1;
L=zeros(w,w);

for k=1:n+1
    V=1;
    for j=1:n+1
        if k~=j
            V=conv(V,poly(X(j)))/(X(k)-X(j));
        end
    end
    L(k,:)=V;
end

C=Y*L;
```

Figure 5-12.

As an example we find the Lagrange interpolating polynomial that passes through the points (2,3), (4,5), (6,5), (7,6), (8,8), (9,7).

We will simply use the following MATLAB syntax:

```
>> [F, L] = lagrange([2 4 6 7 8 9],[3 5 5 6 8 7])
```

C =

-0.0185 0.4857-4.8125 22.2143-46.6690 38.8000

L =

-0.0006 0.0202 -0.2708 1.7798 -5.7286 7.2000
0.0042 -0.1333 1.6458 -9.6667 26.3500 -25.2000
-0.0208 0.6250 -7.1458 38.3750 -94.8333 84.0000
0.0333 -0.9667 10.6667 -55.3333 132.8000 -115.2000
-0.0208 0.5833 -6.2292 31.4167 -73.7500 63.0000
0.0048 -0.1286 1.3333 -6.5714 15.1619 -12.8000

We can obtain the symbolic form of the polynomial whose coefficients are given by the vector C by using the following MATLAB command:

```
>> pretty(poly2sym(C))
```

```
  31    5    1093731338075689   4    77   3    311   2    19601
- ---- x   + ----------------- x   - -- x   + --- x   - ----- x + 194/5
  1680       2251799813685248        16        14         420
```

Newton Polynomial Interpolation

The Newton interpolating polynomial that passes through the $N+1$ points $(x_k, y_k) = (x_k, f(x_k))$, $k = 0, 1, ..., N$, is defined as follows:

$$P(x) = d_{0,0} + d_{1,1}(x - x_0) + d_{2,2}(x - x_0)(x - x_1) + \cdots + d_{N,N}(x - x_0)(x - x_1) \cdots (x - x_{N-1})$$

where:

$$d_{k,j} = y_k \quad d_{k,j} = \frac{d_{k,j-1} - d_{k-1,j-1}}{x_k - d_{k-1}}.$$

Obtaining the coefficients C of the interpolating polynomial and the divided difference table D is easily done via the M-file shown in Figure 5-13.

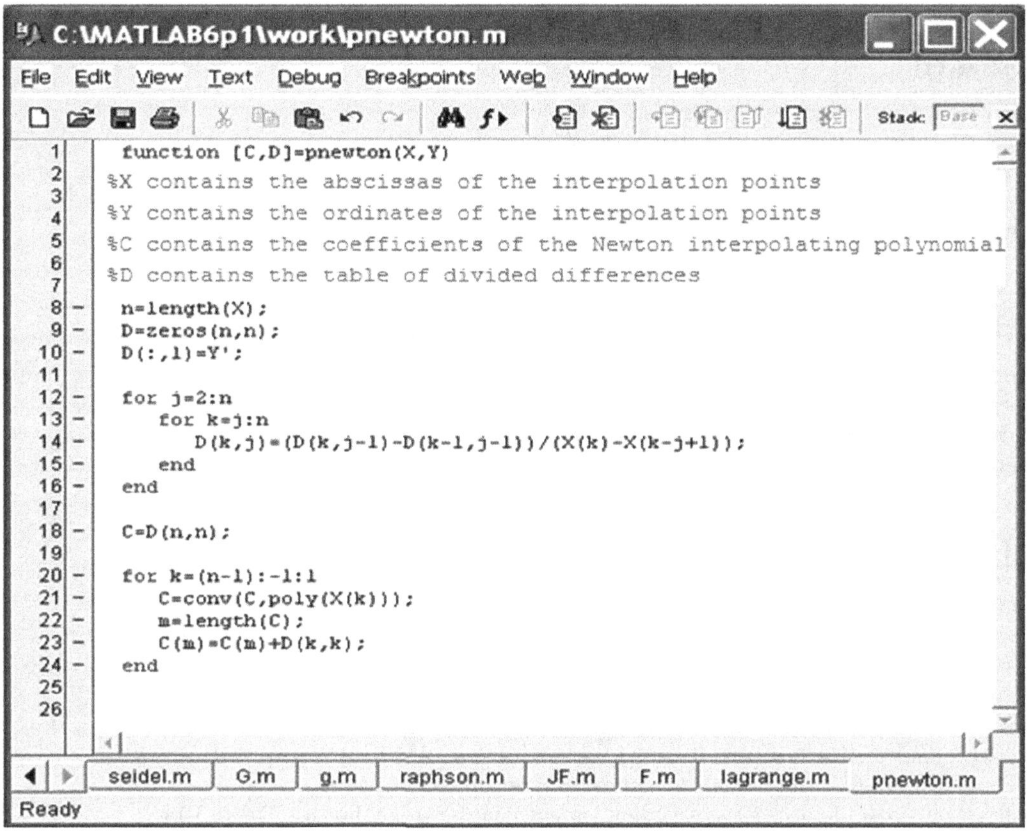

```
🕹 C:\MATLAB6p1\work\pnewton.m                               _ □ ✗

File  Edit  View  Text  Debug  Breakpoints  Web  Window  Help

 □ 🖙 🖫 🎒   ⅍ 🗈 🛱 ㄅ ⌐ | 🗛 ƒ▸ | 🗐 🗗 | 🕀 🕀 🗐 🛄 🗐 | Stack Base ✗

  1        function [C,D]=pnewton(X,Y)
  2       %X contains the abscissas of the interpolation points
  3
  4       %Y contains the ordinates of the interpolation points
  5       %C contains the coefficients of the Newton interpolating polynomial
  6
  7       %D contains the table of divided differences
  8 -      n=length(X);
  9 -      D=zeros(n,n);
 10 -      D(:,1)=Y';
 11
 12 -      for j=2:n
 13 -         for k=j:n
 14 -            D(k,j)=(D(k,j-1)-D(k-1,j-1))/(X(k)-X(k-j+1));
 15 -         end
 16 -      end
 17
 18 -      C=D(n,n);
 19
 20 -      for k=(n-1):-1:1
 21 -         C=conv(C,poly(X(k)));
 22 -         m=length(C);
 23 -         C(m)=C(m)+D(k,k);
 24 -      end
 25
 26

 ◀│▶   seidel.m   G.m   g.m   raphson.m   JF.m   F.m   lagrange.m   pnewton.m
 Ready
```

Figure 5-13.

As an example we apply Newton's method to the same interpolation problem solved by the Lagrange method in the previous section. We will use the following MATLAB syntax:

`>> [C, D] = pnewton([2 4 6 7 8 9],[3 5 5 6 8 7])`

$C =$

$-0.0185 \ 0.4857 - 4.8125 \ 22.2143 - 46.6690 \ 38.8000$

$D =$

3.0000	0	0	0	0	0
5.0000	1.0000	0	0	0	0
5.0000	0	- 0.2500	0	0	0
6.0000	1.0000	0.3333	0.1167	0	0
8.0000	2.0000	0.5000	0.0417	- 0.0125	0
7.0000	- 1.0000	- 1.5000	- 0.6667	- 0.1417	- 0.0185

The interpolating polynomial in symbolic form is calculated as follows:

```
>> pretty(poly2sym(C))
```

$$-\frac{31}{1680}\,x^5 + \frac{5}{35}\,x^4 - \frac{17}{16}\,x^3 + \frac{77}{14}\,x^2 - \frac{311}{420}\,x + \frac{19601}{...}\,x + 194/5$$

Observe that the results obtained by both interpolation methods are similar.

Numerical Derivation Methods

There are various different techniques available for numerical derivation. These are of great importance when developing algorithms to solve problems involving ordinary or partial differential equations.

Among the most common methods for numerical derivation are derivation using limits, derivation using extrapolation and derivation using interpolation on N-1 nodes.

Numerical Derivation via Limits

This method consists in building a sequence of numerical approximations to $f'(x)$ via the generated sequence:

$$f'(x) \approx D_k = \frac{f(x + 10^{-k}h) - f(x - 10^{-k}h)}{2(10^{-k}h)}.$$

The iterations continue until
$|D_{n+1} - D_n| \geq |D_n - D_{n-1}|$ or $|D_n - D_{n-1}| <$ tolerance. This approach approximates $f'(x)$ by D_n.
The algorithm to obtain the derivative D is easily implemented by the M-file shown in Figure 5-14.

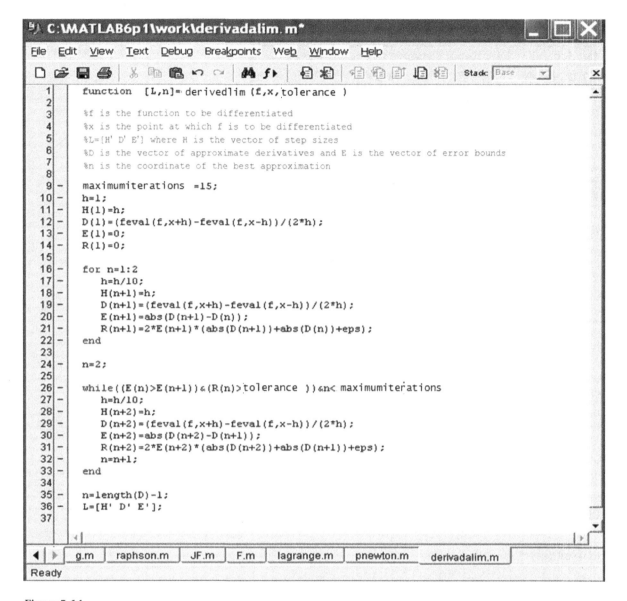

```
C:\MATLAB6p1\work\derivadalim.m*
File  Edit  View  Text  Debug  Breakpoints  Web  Window  Help
                                                          Stack Base

1    function  [L,n]= derivedlim (f,x,tolerance )
2
3    %f is the function to be differentiated
4    %x is the point at which f is to be differentiated
5    %L=[H' D' E'] where H is the vector of step sizes
6    %D is the vector of approximate derivatives and E is the vector of error bounds
7    %n is the coordinate of the best approximation
8
9    maximumiterations =15;
10   h=1;
11   H(1)=h;
12   D(1)=(feval(f,x+h)-feval(f,x-h))/(2*h);
13   E(1)=0;
14   R(1)=0;
15
16   for n=1:2
17       h=h/10;
18       H(n+1)=h;
19       D(n+1)=(feval(f,x+h)-feval(f,x-h))/(2*h);
20       E(n+1)=abs(D(n+1)-D(n));
21       R(n+1)=2*E(n+1)*(abs(D(n+1))+abs(D(n))+eps);
22   end
23
24   n=2;
25
26   while((E(n)>E(n+1))&(R(n)>tolerance ))&n< maximumiterations
27       h=h/10;
28       H(n+2)=h;
29       D(n+2)=(feval(f,x+h)-feval(f,x-h))/(2*h);
30       E(n+2)=abs(D(n+2)-D(n+1));
31       R(n+2)=2*E(n+2)*(abs(D(n+2))+abs(D(n+1))+eps);
32       n=n+1;
33   end
34
35   n=length(D)-1;
36   L=[H' D' E'];
37
```

g.m   raphson.m   JF.m   F.m   lagrange.m   pnewton.m   derivadalim.m

Ready

Figure 5-14.

As an example, we approximate the derivative of the function:

$$f(x) = \sin\left(\cos\left(\frac{1}{x}\right)\right)$$

at the point $\dfrac{1-\sqrt{5}}{2}$.

To begin we define the function *f* in an M-file named funcion (see Figure 5-15). (Note: we use funcion rather than function here since the latter is a protected term in MATLAB.)

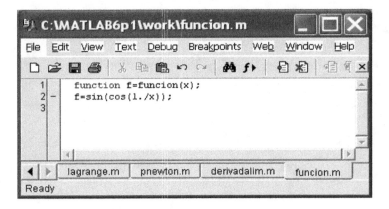

Figure 5-15.

The derivative is then given by the following MATLAB command:

```
>> [L, n] = derivedlim ('funcion', (1-sqrt (5)) / 2,0.01)
```

```
L =
1.0000 - 0.7448 0
0.1000 - 2.6045 1.8598
0.0100 - 2.6122 0.0077
0.0010 - 2.6122 0.0000
0.0001 - 2.6122 0.0000

n =
4
```

Thus we see that the approximate derivative is – 2.6122, which can be checked as follows:

```
>> f = diff ('sin (cos (x))')
```

```
f =
```

```
cos (cos (x)) * sin (x) / x ^ 2
```

```
>> subs (f, (1-sqrt (5)) / 2).
```

```
ans =

  -2.6122
```

Richardson's Extrapolation Method

This method involves building numerical approximations to $f'(x)$ via the construction of a table of values $D(j, k)$ with $k \le j$ that yield a final solution to the derivative $f'(x) = D(n, n)$. The values $D(j, k)$ form a lower triangular matrix, the first column of which is defined as:

$$D(j,1) = \frac{f(x+2^{-j}h) - f(x-2^{-j}h)}{2^{-j+1}h}$$

and the remaining elements are defined by:

$$D(j,k) = D(j,k-1) + \frac{D(j,k-1) - D(j-1,k-1)}{4^k - 1} \quad (2 \le k \le j)$$

The corresponding algorithm for D is implemented by the M-file shown in Figure 5-16.

Figure 5-16.

As an example, we approximate the derivative of the function:

$$f(x) = \sin\left(\cos\left(\frac{1}{x}\right)\right)$$

at the point $\dfrac{1-\sqrt{5}}{2}$.

As the M-file that defines the function f has already been defined in the previous section, we can find the approximate derivative using the MATLAB syntax:

```
>> [D, relativeerror, absoluteerror, n] = richardson ('funcion',
(1-sqrt(5))/2,0.001,0.001)
```

D =

```
-0.7448        0        0        0        0        0
-1.1335 - 1.2631 0 0 0 0
-2.3716 - 2.7843 - 2.8857 0 0 0
-2.5947 - 2.6691 - 2.6614 - 2.6578 0 0
-2.6107 - 2.6160 - 2.6125 - 2.6117 - 2.6115 0
-2.6120 - 2.6124 - 2.6122 - 2.6122 - 2.6122 - 2.6122
```

relativeerror =

   6. 9003e-004

absoluteerror =

   2. 6419e-004

n =
6

Thus we get the same result as before when we used the limit method.

Derivation Using Interpolation (n + 1 nodes)

This method consists in building the Newton interpolating polynomial of degree N:

$$P(x) = a_0 + a_1(x - x_0) + a_2(x - x_0)(x - x_1) + \cdots + a_N(x - x_0)(x - x_1)\cdots(x - x_{N-1})$$

and numerically approximating $f'(x_0)$ by $P'(x_0)$.

The algorithm for the derivative D is easily implemented by the M-file shown in Figure 5-17.

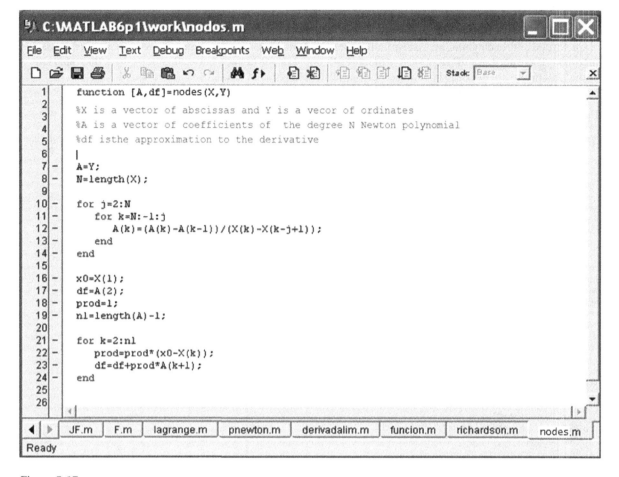

Figure 5-17.

As an example, we approximate the derivative of the function:

$$f(x) = \sin\left(\cos\left(\frac{1}{x}\right)\right)$$

at the point $\frac{1-\sqrt{5}}{2}$.

As the M-file that defines the function *f* has already been constructed in the previous section, we can calculate the approximate derivative using the MATLAB command:

```
>> [A, df] = nodes([2 4 6 7 8 9],[3 5 5 6 8 7])

A =

3.0000 1.0000 - 0.2500 0.1167 - 0.0125 - 0.0185

df =
-1.4952
```

Numerical Integration Methods

Given the difficulty of obtaining an exact primitive for many functions, numerical integration methods are especially important. There are many different ways to numerically approximate definite integrals, among them the trapezium method, Simpson's method and Romberg's method (all implemented in MATLAB's Basic module).

The Trapezium Method

The trapezium method for numerical integration has two variants: the trapezoidal rule and the recursive trapezoidal rule. The trapezoidal rule approximates the definite integral of the function $f(x)$ between a and b as follows:

$$\int_a^b f(x)dx \approx \frac{h}{2}(f(a)+f(b))+h\sum_{k=1}^{M-1}f(x_k)$$

calculating $f(x)$ at equidistant points $x_k = a+kh$, $k=0,1,...,M$ where $x_0=a$ and $x_M=b$.

The trapezoidal rule is implemented by the M-file shown in Figure 5-18.

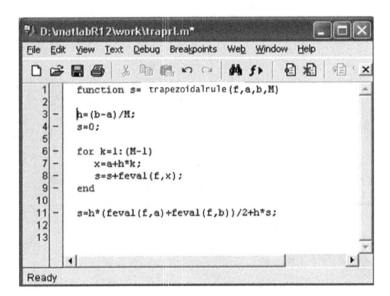

```
function s= trapezoidalrule(f,a,b,M)

h=(b-a)/M;
s=0;

for k=1:(M-1)
    x=a+h*k;
    s=s+feval(f,x);
end

s=h*(feval(f,a)+feval(f,b))/2+h*s;
```

Figure 5-18.

The *recursive trapezoidal rule* considers the points $x_k = a+kh$, $k=0,1,...,M$, where $x_0=a$ and $x_M=b$, dividing the interval $[a, b]$ into $2J=M$ subintervals of the same size $h=(b-a)/2J$. We then consider the following recursive formula:

$$T(0)=\frac{h}{2}(f(a)+f(b))$$

$$T(J)=\frac{T(J-1)}{2}+h\sum_{k=1}^{M}f(x_{2k-1})$$

and the integral of the function $f(x)$ between a and b can be calculated as:

$$\int_a^b f(x)dx \approx \frac{h}{2}\sum_{k=1}^{2^J}(f(x_k)+f(x_{k-1}))$$

using the trapezoidal rule as the number of sub-intervals $[a, b]$ increases, taking at the J-th iteration a set of $2J+1$ equally spaced points.

The recursive trapezoidal rule is implemented via the M-file shown in Figure 5-19.

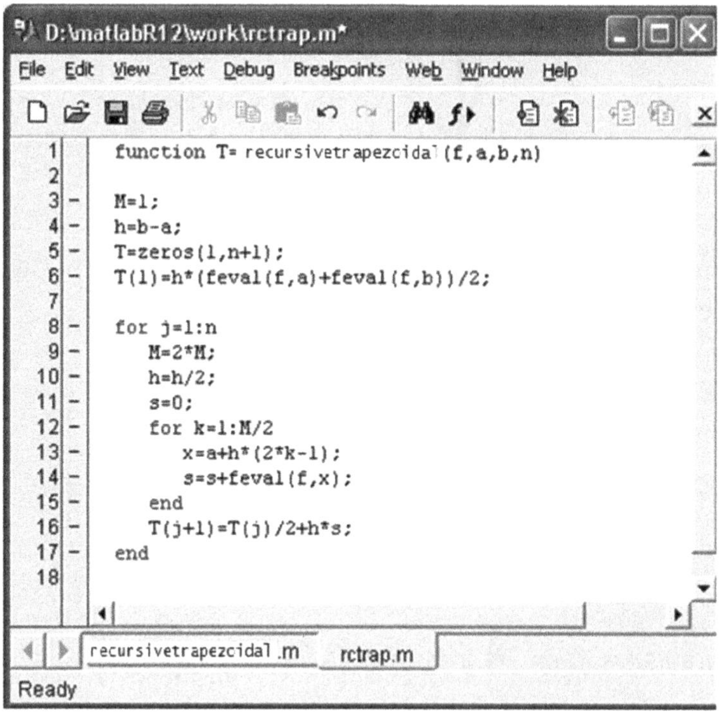
D:\matlabR12\work\rctrap.m*
File Edit View Text Debug Breakpoints Web Window Help
function T= recursivetrapezcidal (f,a,b,n)
M=1;
h=b-a;
T=zeros(1,n+1);
T(1)=h*(feval(f,a)+feval(f,b))/2;
for j=1:n
M=2*M;
h=h/2;
s=0;
for k=1:M/2
x=a+h*(2*k-1);
s=s+feval(f,x);
end
T(j+1)=T(j)/2+h*s;
end
recursivetrapezcidal.m
rctrap.m
Ready

Figure 5-19.

As an example, we calculate the following integral using 100 iterations of the recursive trapezoidal rule:

$$\int_0^2 \frac{1}{x^2 + \frac{1}{10}} dx.$$

We start by defining the integrand by means of the M-file *integrand1.m* shown in Figure 5-20.

Figure 5-20.

We then calculate the requested integral as follows:

```
>> recursivetrapezoidal('integrand1',0,2,14)

ans =

Columns 1 through 4

10.24390243902439    6.03104212860310    4.65685845031979    4.47367657743630

Columns 5 through 8

4.47109102437123 4.47132194954670 4.47138003053334 4.47139455324593

Columns 9 through 12

4.47139818407829 4.47139909179602 4.47139931872606 4.47139937545860

Columns 13 through 15

4.47139938964175 4.47139939318754 4.47139939407398
```

This shows that after 14 iterations an accurate value for the integral is 4.47139939407398.
We calculate the same integral using the trapezoidal rule, using M = 14, using the following MATLAB command:

```
>> trapezoidalrule('integrand1',0,2,14)

ans =
4.47100414648074
```

The result is now the less accurate 4.47100414648074.

Simpson's Method

Simpson's method for numerical integration is generally considered in two variants: the simple Simpson's rule and the composite Simpson's rule.

Simpson's simple approximation of the definite integral of the function $f(x)$ between the points a and b is the following:

$$\int_a^b f(x)dx \approx \frac{h}{3}(f(a)+f(b)+4f(c))\; c=\frac{a+b}{2}$$

This can be implemented using the M-file shown in Figure 5-21.

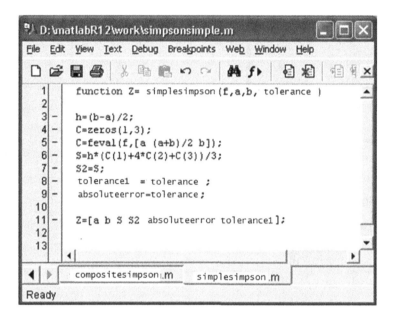

Figure 5-21.

The *composite Simpson's rule* approximates the definite integral of the function $f(x)$ between points a and b as follows:

$$\int_a^b f(x)dx \approx \frac{h}{3}(f(a)+f(b))+\frac{2h}{3}\sum_{k=1}^{M-1}f(x_{2k})+\frac{4h}{3}\sum_{k=1}^{M}f(x_{2k-1})$$

calculating $f(x)$ at equidistant points $x_k = a+kh$, k= 0, 1,..., 2M, where $x_0=a$ and $x_{2M}=b$.
The composite Simpson's rule is implemented using the M-file shown in Figure 5-22.

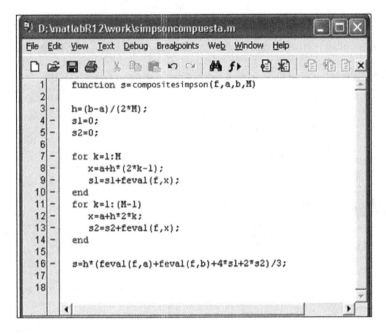

```
function s=compositesimpson(f,a,b,M)

h=(b-a)/(2*M);
s1=0;
s2=0;

for k=1:M
    x=a+h*(2*k-1);
    s1=s1+feval(f,x);
end
for k=1:(M-1)
    x=a+h*2*k;
    s2=s2+feval(f,x);
end

s=h*(feval(f,a)+feval(f,b)+4*s1+2*s2)/3;
```

Figure 5-22.

As an example, we calculate the following integral by the composite Simpson's rule taking M = 14:

$$\int_0^2 \frac{1}{x^2 + \frac{1}{10}}\, dx.$$

We use the following syntax:

```
>>compositesimpson('integrand1',0,2,14)

ans =

   4.47139628125498
```

Next we calculate the same integral using the simple Simpson's rule:

```
>> Z=simplesimpson('integrand2',0,2,0.0001)

Z =

Columns 1 through 4

0 2.00000000000000 4.62675535846268 4.62675535846268

Columns 5 through 6

0.00010000000000 0.00010000000000
```

As we see, the simple Simpson's rule is less accurate than the composite rule.

In this case, we have previously defined the integrand in the M-file named *integrand2.m* (see Figure 5-23).

```
function F=integrand2(x);
F=1./(x.^2+1/10);
```

Figure 5-23.

Ordinary Differential Equations

Obtaining exact solutions of ordinary differential equations is not a simple task. There are a number of different methods for obtaining approximate solutions of ordinary differential equations. These numerical methods include, among others, Euler's method, Heun's method, the Taylor series method, the Runge–Kutta method (implemented in MATLAB's Basic module), the Adams–Bashforth–Moulton method, Milne's method and Hamming's method.

Euler's Method

Suppose we want to solve the differential equation $y' = f(t, y)$, $y(a) = y_0$, on the interval $[a, b]$. We divide the interval $[a, b]$ into M subintervals of the same size using the partition given by the points $t_k = a + kh$, $k = 0, 1, ..., M$, $h = (b-a)/M$. Euler's method then finds the solution of the differential equation iteratively by calculating $y_{k+1} = y_k + hf(t_k, y_k)$, $k = 0, 1, ..., M-1$.

Euler's method is implemented using the M-file shown in Figure 5-24.

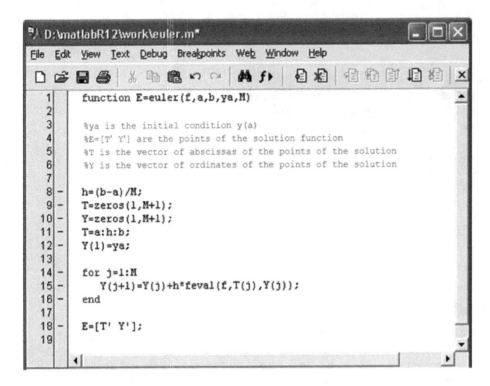

```
function E=euler(f,a,b,ya,M)

%ya is the initial condition y(a)
%E=[T' Y'] are the points of the solution function
%T is the vector of abscissas of the points of the solution
%Y is the vector of ordinates of the points of the solution

h=(b-a)/M;
T=zeros(1,M+1);
Y=zeros(1,M+1);
T=a:h:b;
Y(1)=ya;

for j=1:M
    Y(j+1)=Y(j)+h*feval(f,T(j),Y(j));
end

E=[T' Y'];
```

Figure 5-24.

Heun's Method

Suppose we want to solve the differential equation $y' = f(t, y)$, $y(a) = y_0$, on the interval $[a, b]$. We divide the interval $[a, b]$ into M subintervals of the same size using the partition given by the points $t_k = a + kh$, $k = 0, 1, ..., M$, $h = (b-a)/M$. Heun's method then finds the solution of the differential equation iteratively by calculating $y_{k+1} = y_k + h(f(t_k, y_k) + f(t_{k+1}, y_k + f(t_k, y_k))) / 2$, $k = 0, 1, ..., M-1$.

Heun's method is implemented using the M-file shown in Figure 5-25.

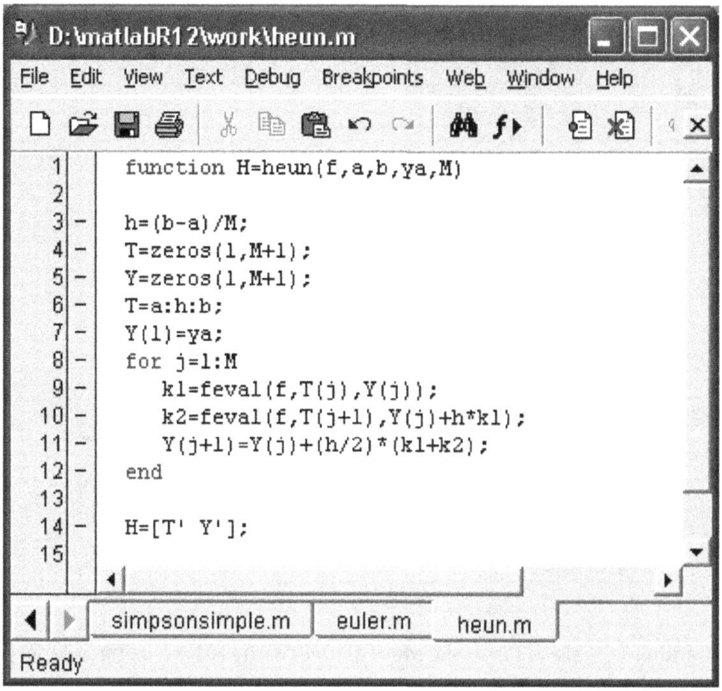

Figure 5-25.

The Taylor Series Method

Suppose we want to solve the differential equation $y' = f(t, y)$, $y(a) = y_0$, on the interval $[a, b]$. We divide the interval $[a, b]$ into M subintervals of the same size using the partition given by the points $t_k = a + kh$, k = 0,1,..., M, $h = (b-a)/M$. The Taylor series method (let us consider here the method to order 4) finds a solution to the differential equation by evaluating y', y'', y''' and y'''' to give the 4th order Taylor series for y at each partition point.

The Taylor series method is implemented using the M-file shown in Figure 5-26.

D:\matlabR12\work\taylor.m
File Edit View Text Debug Breakpoints Web Window Help
function T4=taylor(df,a,b,ya,M)

% df=[y' y'' y''' y''''] is the string 'df'
% T4=[T' Y']

h=(b-a)/M;
T=zeros(1,M+1);
Y=zeros(1,M+1);
T=a:h:b;
Y(1)=ya;

for j=1:M
D=feval(df,T(j),Y(j));
Y(j+1)=Y(j)+h*(D(1)+h*(D(2)/2+h*(D(3)/6+h*D(4)/24)));
end

T4=[T' Y'];

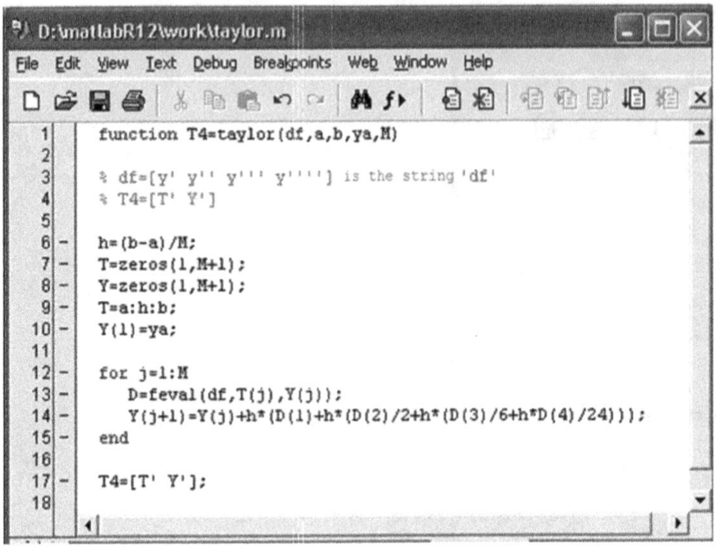

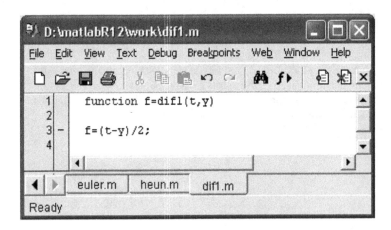

Figure 5-26.

As an example we solve the differential equation $y'(t) = (t - y) / 2$ on the interval [0,3], with $y(0) = 1$, using Euler's method, Heun's method and by the Taylor series method.

We will begin by defining the function $f(t, y)$ via the M-file shown in Figure 5-27.

D:\matlabR12\work\dif1.m
File Edit View Text Debug Breakpoints Web Window Help
function f=dif1(t,y)

f=(t-y)/2;
euler.m
heun.m
dif1.m
Ready

Figure 5-27.

The solution of the equation using Euler's method in 100 steps is calculated as follows:

```
>> E = euler('dif1',0,3,1,100)

E =

0 1.00000000000000
0.03000000000000 0.98500000000000
```

```
0.06000000000000  0.97067500000000
0.09000000000000  0.95701487500000
0.12000000000000  0.94400965187500
0.15000000000000  0.93164950709688
0.18000000000000  0.91992476449042
.
.
.
2.85000000000000  1.56377799005910
2.88000000000000  1.58307132020821
2.91000000000000  1.60252525040509
2.94000000000000  1.62213737164901
2.97000000000000  1.64190531107428
3.00000000000000  1.66182673140816
```

This solution can be graphed as follows (see Figure 5-28):

>> plot (E (:,2))

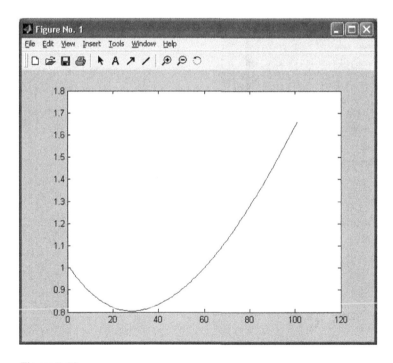

Figure 5-28.

The solution of the equation by Heun's method in 100 steps is calculated as follows:

```
>> H = heun('dif1',0,3,1,100)
H =
0 1.00000000000000
0.03000000000000  0.98533750000000
0.06000000000000  0.97133991296875
0.09000000000000  0.95799734001443
0.12000000000000  0.94530002961496
.
.
.
2.88000000000000  1.59082209379464
2.91000000000000  1.61023972987327
2.94000000000000  1.62981491089478
2.97000000000000  1.64954529140884
3.00000000000000  1.66942856088299
```

The solution using the Taylor series method requires the previously defined function $df = [y'\ y''\ y'''\ y'''']$ using the M-file shown in Figure 5-29.

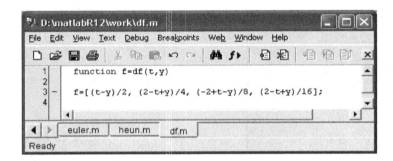

Figure 5-29.

The differential equation is solved by the Taylor series method via the command:

```
>> T = taylor('df',0,3,1,100)
```

```
T =
0 1.00000000000000
0.03000000000000  0.98533581882813
0.06000000000000  0.97133660068283
0.09000000000000  0.95799244555443
0.12000000000000  0.94529360082516
.
.
.
2.88000000000000  1.59078327648360
2.91000000000000  1.61020109213866
2.94000000000000  1.62977645599332
2.97000000000000  1.64950702246046
3.00000000000000  1.66939048087422
```

EXERCISE 5-1

Solve the following non-linear equation using the fixed point iterative method:

$$x = \cos(\sin(x)).$$

We will start by finding an approximate solution to the equation, which we will use as the initial value p_0. To do this we show the x axis and the curve y=x-cos(sin(x)) on the same graph (Figure 5-30) by using the following command:

```
>> fplot ([x-cos (sin (x)), 0], [- 2, 2])
```

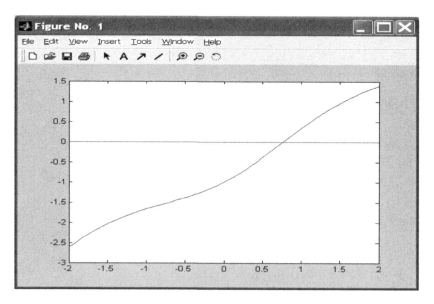

Figure 5-30.

The graph indicates that there is a solution close to $x = 1$, which is the value that we shall take as our initial approximation to the solution, i.e. $p_0 = 1$. If we consider a tolerance of 0.0001 for a maximum number of 100 iterations, we can solve the problem once we have defined the function $g(x) = \cos(\sin(x))$ via the M-file $g91.m$ shown in Figure 5-31.

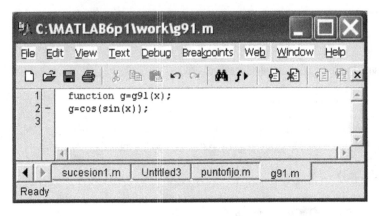

Figure 5-31.

We can now solve the equation using the MATLAB command:

>> *[k, p, absoluteerror, P]=fixedpoint('g91',1,0.0001,1000)*

k =

   13

p =

   0.7682

absoluteerror =

  6. 3361e-005

P =

1.0000
0.6664
0.8150
0.7467
0.7781
0.7636
0.7703
0.7672
0.7686
0.7680
0.7683
0.7681
0.7682

The solution is $x = 0.7682$, which has been found in 13 iterations with an absolute error of 6.3361e- 005. Thus, the convergence to the solution is particularly fast.

EXERCISE 5-2

Using Newton's method calculate the root of the equation $x^3 - 10x^2 + 29x - 20 = 0$ close to the point $x = 7$ with an accuracy of 0.00005. Repeat the same calculation but with an accuracy of 0.0005.

We define the function $f(x) = x^3 - 10x^2 + 29x - 20$ and its derivative via the M-files named *f302.m* and *f303.m* shown in Figures 5-32 and 5-33.

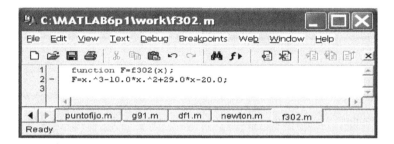

Figure 5-32.

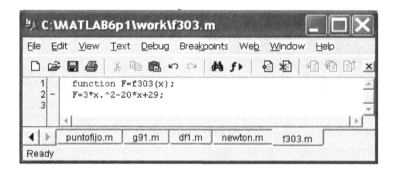

Figure 5-33.

To run the program that solves the equation, type:

```
>> [x, it]=newton('f302','f303',7,.00005)

x =

5.0000

it =

6
```

In 6 iterations and with an accuracy of 0.00005 the solution $x = 5$ has been obtained. In 5 iterations and with an accuracy of 0.0005 we get the solution $x = 5.0002$:

```
>> [x, it] = newton('f302','f303',7,.0005)

x =

5.0002

it =

5
```

EXERCISE 5-3

Write a program that calculates a root with multiplicity 2 of the equation $(e^{-x} - x)^2 = 0$ close to the point $x = -2$ to an accuracy of 0.00005.

We define the function $f(x)=(e^x - x)^2$ and its derivative via the M-files *f304.m* and *f305.m* shown in Figures 5-34 and 5-35:

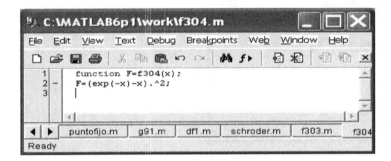

Figure 5-34.

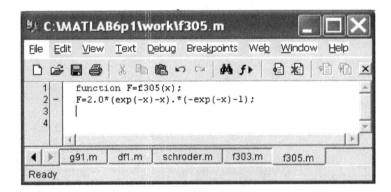

Figure 5-35.

We solve the equation using Schröder's method. To run the program we enter the command:

```
>> [x,it]=schroder('f304','f305',2,-2,.00005)

x =

0.5671

it =

5
```

In 5 iterations we have found the solution $x = 0.56715$.

EXERCISE 5-4

Approximate the derivative of the function

$$f(x) = \tan\left(\cos\left(\frac{\sqrt{5} + \sin(x)}{1 + x^2}\right)\right)$$

at the point $\dfrac{1 - \sqrt{5}}{3}$.

To begin we define the function *f* in the M-file *funcion1.m* shown in Figure 5-36.

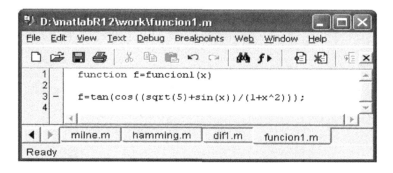

Figure 5-36.

The derivative can be found using the method of numerical derivation with an accuracy of 0.0001 via the following MATLAB command:

```
>> [L, n] = derivedlim ('funcion1', (1 + sqrt (5)) / 3,0.0001)

L =

1.00000000000000 0.94450896913313 0
0.10000000000000 1.22912035588668 0.28461138675355
0.01000000000000 1.22860294102802 0.00051741485866
```

0.00100000000000 1.22859747858110 0.00000546244691
0.00010000000000 1.22859742392997 0.00000005465113

n =

4

We see that the value of the derivative is approximated by 1.22859742392997.

Using Richardson's method, the derivative is calculated as follows:

>> [D, absoluteerror, relativeerror, n] = ('funcion1' richardson,(1+sqrt(5))/3,0.0001,0.0001)

D =

Columns 1 through 4

0.94450896913313	0	0	0
1.22047776163545	1.31246735913623	0	0
1.23085024935646	1.23430774526347	1.22909710433862	0
1.22938849854454	1.22890124827389	1.22854081514126	1.22853198515400
1.22880865382036	1.22861537224563	1.22859631384374	1.22859719477553

Column 5

0
0
0
0
1.22859745049954

absoluteerror =

6. 546534553897310e-005

relativeerror =

5. 328603742973844e-005

n =
5

EXERCISE 5-5

Approximate the following integral:

$$\int_{1}^{\frac{2\pi}{3}} \tan\left(\cos\left(\frac{\sqrt{5}+\sin(x)}{1+x^2}\right)\right) dx.$$

We can use the composite Simpson's rule with M=100 using the following command:

```
>> s = compositesimpson('function1',1,2*pi/3,100)
```

s =

0.68600990924332

If we use the trapezoidal rule instead, we get the following result:

```
>> s = trapezoidalrule('function1',1,2*pi/3,100)
```

s =

0.68600381840334

EXERCISE 5-6

Find an approximate solution of the following differential equation in the interval [0, 0.8]:

$$y' = t^2 + y^2 \quad y(0) = 1.$$

We start by defining the function $f(t, y)$ via the M-file in Figure 5-37.

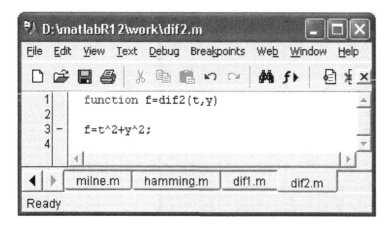

Figure 5-37.

We then solve the differential equation by Euler's method, dividing the interval into 20 subintervals using the following command:

```
>> E = euler('dif2',0,0.8,1,20)
```

```
E =
0  1.00000000000000
0.04000000000000  1.04000000000000
0.08000000000000  1.08332800000000
0.12000000000000  1.13052798222336
0.16000000000000  1.18222772296696
0.20000000000000  1.23915821852503
0.24000000000000  1.30217874214655
0.28000000000000  1.37230952120649
0.32000000000000  1.45077485808625
0.36000000000000  1.53906076564045
0.40000000000000  1.63899308725380
0.44000000000000  1.75284502085643
0.48000000000000  1.88348764754208
0.52000000000000  2.03460467627982
0.56000000000000  2.21100532382941
0.60000000000000  2.41909110550949
0.64000000000000  2.66757117657970
0.68000000000000  2.96859261586445
0.72000000000000  3.33959030062305
0.76000000000000  3.80644083566367
0.80000000000000  4.40910450907999
```

The solution can be graphed as follows (see Figure 5-38):

```
>> plot (E (:,2))
```

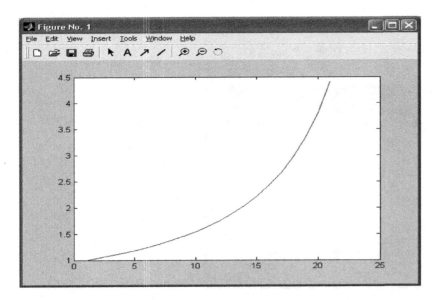

Figure 5-38.

Get the eBook for only $10!

Now you can take the weightless companion with you anywhere, anytime. Your purchase of this book entitles you to 3 electronic versions for only $10.

This Apress title will prove so indispensible that you'll want to carry it with you everywhere, which is why we are offering the eBook in **3 formats** for only $10 if you have already purchased the print book.

Convenient and fully searchable, the PDF version enables you to easily find and copy code—or perform examples by quickly toggling between instructions and applications. The MOBI format is ideal for your Kindle, while the ePUB can be utilized on a variety of mobile devices.

Go to www.apress.com/promo/tendollars to purchase your companion eBook.

Apress®
THE EXPERT'S VOICE™